Use R!

Use R!

This series of inexpensive and focused books on R will publish shorter books aimed at practitioners. Books can discuss the use of R in a particular subject area (e.g., epidemiology, econometrics, psychometrics) or as it relates to statistical topics (e.g., missing data, longitudinal data). In most cases, books will combine LaTeX and R so that the code for figures and tables can be put on a website. Authors should assume a background as supplied by Dalgaard's Introductory Statistics with R or other introductory books so that each book does not repeat basic material.

More information about this series at http://www.springer.com/series/6991

Ron Wehrens

Chemometrics with R

Multivariate Data Analysis in the Natural and Life Sciences

Second Edition

 Springer

Ron Wehrens
Wageningen University
Wageningen, The Netherlands

ISSN 2197-5736   ISSN 2197-5744   (electronic)
Use R!
ISBN 978-3-662-62026-7   ISBN 978-3-662-62027-4   (eBook)
https://doi.org/10.1007/978-3-662-62027-4

This Springer imprint is published by the registered company Springer-Verlag GmbH, DE part of Springer Nature.
The registered company address is: Heidelberger Platz 3, 14197 Berlin, Germany

For Odilia, Chris and Luc

Preface to the Second Edition

Eight years after the appearance of the first edition of this book, the R ecosystem has evolved significantly. The number of R users has continued to grow, and so has the number of R packages. The latter not only can be found at the two main R repositories, CRAN and Bioconductor, where elementary quality control checks are being applied to ascertain that the packages are at least semantically correct and provide a minimum of support, but also at other platforms such as github. Installation of an R package is as easy as can be, probably one of the main reasons for the huge success that the language is experiencing. At the same time, this presents the user with a difficult problem: where should I look? What should I use? In keeping with the aims formulated in the first edition, this second edition presents an overview of techniques common in chemometrics, and R packages implementing them. I have tried to remain as close as possible to elementary packages, i.e., packages that have been designed for one particular purpose and do this pretty well. All of them are from CRAN or Bioconductor.

Maybe somewhat ironically, the R package **ChemometricsWithR**, accompanying this book, will no longer be hosted on CRAN. Due to package size restrictions, the package accompanying the first edition was forced to be split into two, the data part transferring into a separate package, **ChemometricsWithRData**. For this new edition, hosting everything on my own github repository has made it possible to reunite the two packages, making life easier for the reader. Installing the **ChemometricsWithR** package can be done as follows:

```
> install.packages("remotes")
> library(remotes)
> install_github("rwehrens/ChemometricsWithR")
```

All code in the book is also present in the demo subdirectory of the package, and can be run directly: each chapter constitutes one demo. These demos make copying-and-pasting from the book unnecessary.

Many of the packages mentioned in the first edition, and indeed the R language itself, have changed over the years—some packages are no longer available, are superseded by other packages, or have seen other changes. Repairing all the resulting inconsistencies with the first edition was the main motivation to revisit and revise the book. Of course, the opportunity to reorganize and extend the material was taken. For one thing, figures are now in color, which not only leads to a livelier reading experience but in many cases increases the information content of the plots. Multivariate classification methods, PLSDA and PCDA, now are treated in Chap. 8. Some paragraphs, like the ones on global optimization methods (simulated annealing and genetic algorithms), have been rewritten quite extensively, now focusing more on CRAN R packages than on the educational code that was provided in the first edition. Chapter 11 has been extended with paragraphs on missing values, multivariate process monitoring, biomarker identification and batch correction.

The biggest change, however, has taken place under the hood. In this edition I have used the knitr package by Yihui Xie (Xie 2014, 2015, 2018) a package providing seamless integration of R code and LaTeX that has completely changed my way of working with R. This has helped enormously in straightening out the inconsistencies in the code, and in avoiding errors such as the mismatch between Fig. 3.6 and related text that were present in the first edition. In many cases methods depend on a random initiation, and although random seeds were used to guarantee reproducible results when writing the book, these random seeds are not mentioned in the book text. They are, however, included in the code that can be found in the demo subdirectory of the **ChemometricsWithR** package so that running the demos *will* lead to the results in the book.

Finally, I would like to thank the following people for proofreading and providing constructive feedback: Manya Afonso, Jasper Engel, Jos Hageman, Maikel Verouden and Pietro Franceschi.

Arnhem, The Netherlands   Ron Wehrens
2019

Preface to the First Edition

The natural sciences, and the life sciences in particular, have seen a huge increase in the amount and complexity of data being generated with every experiment. It is only some decades ago that scientists were typically measuring single numbers – weights, extinctions, absorbances – usually directly related to compound concentrations. Data analysis came down to estimating univariate regression lines, uncertainties and reproducibilities. Later, more sophisticated equipment generated complete spectra, where the response of the system is wavelength-dependent. Scientists were confronted with the question how to turn these spectra into usable results such as concentrations. Things became more complex after that: chromatographic techniques for separating mixtures were coupled to high-resolution (mass) spectrometers, yielding a data matrix for every sample, often with large numbers of variables in both chromatographic and spectroscopic directions. A set of such samples corresponds to a data cube rather than a matrix. In parallel, rapid developments in biology saw a massive increase in the ratio of variables to objects in that area as well.

As a result, scientists today are faced with the increasingly difficult task to make sense of it all. Although most will have had a basic course in statistics, such a course is unlikely to have covered much multivariate material. In addition, many of the classical concepts have a rough time when applied to the types of data encountered nowadays – the multiple-testing problem is a vivid illustration. Nevertheless, even though data analysis has become a field in itself (or rather: a large number of specialized fields), scientists generating experimental data should know at least some of the ways to interpret their data, if only to be able to ascertain the quality of what they have generated. Cookbook approaches, involving blindly pushing a sequence of buttons in a software package, should be avoided. Sometimes the things that deviate from expected behavior are the most interesting in a data set, rather than unfortunate measurement errors. These deviations can show up at any time point during data analysis, during data preprocessing, modelling, interpretation... Every phase in this pipeline should be carefully executed and results, also at an intermediate stage, should be checked using common sense and prior knowledge.

This also puts restrictions on the software that is being used. It should be sufficiently powerful and flexible to fit complicated models and handle large and complex data sets, and on the other hand should allow the user to exactly follow what is being calculated – black-box software should be avoided if possible. Moreover, the software should allow for reproducible results, something that is hard to achieve with many point-and-click programs: even with a reasonably detailed prescription, different users can often obtain quite different results. R R Development Core Team (2010), with its rapidly expanding user community, nicely fits the bill. It is quickly becoming the most important tool in statistical bioinformatics and related fields. The base system already provides a large array of useful procedures; in particular, the high-quality graphics system should be mentioned. The most important feature, however, is the package system, allowing users to contribute software for their own fields, containing manual pages and examples that are directly executable. The result is that many packages have been contributed by users for specific applications; the examples and the manual pages make it easy to see what is happening.

Purpose of this book.

Something of this philosophy also can be found in the way this book is set up. The aim is to present a broad field of science in an accessible way, mainly using illustrative examples that can be reproduced immediately by the reader. It is written with several goals in mind:

- **An introduction to multivariate analysis**. On an abstract level, this book presents the route from raw data to information. All steps, starting from the data preprocessing and exploratory analysis to the (statistical) validation of the results, are considered. For students or scientists with little experience in handling real data, this provides a general overview that is sometimes hard to get from classical textbooks. The theory is presented as far as necessary to understand the principles of the methods and the focus is on immediate application on data sets, either from real scientific examples, or specifically suited to illustrate characteristics of the analyses.
- **An introduction to R**. For those scientists already working in the fields of bioinformatics, biostatistics and chemometrics but using other software, the book provides an accessible overview on how to perform the most common analyses in R R Development Core Team (2010). Many packages are available on the standard repositories, CRAN[1] and BIOCONDUCTOR[2], but for people unfamiliar with the basics of R the learning curve can be pretty steep – for software, power and complexity are usually correlated. This book is an attempt to provide a more gentle way up.

---

[1] http://cran.r-project.org.

[2] http://www.bioconductor.org.

- **Combining multivariate data analysis and R**. The combination of the previous two goals is especially geared towards university students, at the beginning of their specialization: it is of prime importance to obtain hands-on experience on real data sets. It does take some help to start reading R code - once a certain level has been reached, it becomes more easy. The focus, therefore, is not just on the use of the many packages that are available, but also on showing how the methods are implemented. In many cases, simplified versions of the algorithms are given explicitly in the text, so that the reader is able to follow step-by-step what is happening. It is this insight in (at least the basics of) the techniques that are essential for fruitful application.

The book has been explicitly set up for self-study. The user is encouraged to try out the examples, and to substitute his or her own data as well. If used in a university course, it is possible to keep the classical "teaching" of theory to a minimum; during the lessons, teachers can concentrate on the analysis of real data. There is no substitute for practice.

Prior knowledge.

Some material is assumed to be familiar. Basic statistics, for example, including hypothesis tests, the construction of confidence intervals, analysis of variance and least-squares regression are referred to, but not explained. The same goes for basic matrix algebra. The reader should have some experience in programming in general (variables, variable types, functions, program control, etcetera). It is assumed the reader has installed R, and has a basic working knowledge of R, roughly corresponding to having worked through the excellent "Introduction to R" Venables et al. (2009), which can be found on the CRAN website. In some cases, less mundane functions will receive a bit more attention in the text; examples are the apply and sweep functions. We will only focus on the command-line interface: Windows users may find it easier to perform actions using point-and-click.

*The R package **ChemometricsWithR**.*

With the book comes a package, too: **ChemometricsWithR** contains all data sets and functions used in this book. Installing the package will cause all other packages used in the book to be available as well – an overview of these packages can be found in the Appendix in Part 11.8.5 of the book. In the examples, it is always assumed that the **ChemometricsWithR** package is loaded; where functions or data sets from other packages are used for the first time, this is explicitly mentioned in the text.

More information about the data sets used in the book can be found in the references – no details will be given about the background or interpretation of the measurement techniques.

Acknowledgements.

This book has its origins in a reader for the Chemometrics course at the Radboud University Nijmegen covering exploratory analysis (PCA), clustering (hierarchical methods and k-means), discriminant analysis (LDA, QDA) and multivariate regression (PCR, PLS). Also material from a later course in Pattern Recognition has been included. I am grateful for all the feedback from the students, and especially for the remarks, suggestions and criticisms from my colleagues at the Department of Analytical Chemistry of the Radboud University Nijmegen. I am indebted to Patrick Krooshof and Tom Bloemberg, who have contributed in a major way in developing the material for the courses. Finally, I would like to thank all who have read (parts of) the manuscript and with their suggestions have helped improving it, in particular Tom Bloemberg, Karl Molt, Lionel Blanchet, Pietro Franceschi, and Jan Gerretzen.

Trento   Ron Wehrens
September 2010

Contents

Chapter 1
Introduction

In the last twenty years, the life sciences have seen a dramatic increase in the size and number of data sets. Simple sensing devices in many cases offer real-time data streaming but also more complicated data types such as spectra or images can be acquired at much higher rates then was thought possible not too long ago. At the same time, the complexity of the data that are acquired also increases, in some cases leading to very high numbers of variables—in mass spectrometry it is not uncommon to have tens of thousands of mass peaks; in genomics having millions of single-nucleotide polymorphisms (SNPs) is becoming the rule rather than the exception. "Simply measure everything" is the approach nowadays: rather than focusing on measuring specific predefined characteristics of the sample[1] modern techniques aim at generating a holistic view, sometimes called a "fingerprint". As a result, the analysis of one single sample can easily yield megabytes of data. These (physical) samples typically are complex mixtures and may, e.g., correspond to body fluids of patients and controls, measured with possibly several different spectroscopic techniques; environmental samples (air, water, soil); measurements on different cell cultures or one cell culture under different treatments; industrial samples from process industry, pharmaceutical industry or food industry; samples of competitor products; quality control samples, and many others.

The types of data we will concentrate on are generated by analytical chemical measurement techniques, and are in almost all cases directly related to concentrations or amounts of specific classes of chemicals such as metabolites or proteins. The corresponding research fields are called metabolomics and proteomics, and a host of others with similar characteristics and a similar suffix, collectively known as the "-omics sciences", exist. A well-known example from molecular biology is transcriptomics, focusing on the levels of mRNA obtained by transcription from DNA

---

[1]The word "sample" will be used both for the physical objects on which measurements are performed (the chemical use of the word) and for the current realization of all possible measurements (the statistical use). Which one is meant should be clear from the context.

© Springer-Verlag GmbH Germany, part of Springer Nature 2020
R. Wehrens, *Chemometrics with R*, Use R!,
https://doi.org/10.1007/978-3-662-62027-4_1

strains. Although we do not include any transcriptomics data, many of the techniques treated in this book are directly applicable—in that sense, the characteristics of data of completely different origins can still be comparable.

These data can be analyzed at different levels. The most direct approach is to analyst them as raw data (intensities, spectra, ...), without any prior interpretation other than a suitable pretreatment. Although this has the advantage that it is completely objective, it is usually also more difficult: typically, the number of variables is huge and the interpretability of the statistical models that are generated to describe the data often is low. A more often used strategy is to apply domain knowledge to convert the raw data into more abstract variables such as concentrations, for example by quantifying a set of compounds in a mixture based on a library of pure spectra. The advantage is that the statistical analysis can be performed on the quantities that really matter, and that the models are simpler and easier to validate and interpret. The obvious disadvantage is the dependence on the interpretation step: not always it is easy to decide which compounds are present and in what amounts. Any error at this stage cannot be corrected in later analysis stages.

The extremely rapid development of analytical techniques in biology and chemistry has left data analysis far behind, and as a result the statistical analysis and interpretation of the data has become a major bottleneck in the pipeline from measurement to information. Academic training in multivariate statistics in the life sciences is lagging. Bioinformatics departments are the primary source of scientists with such a background, but bioinformatics is a very broad field covering many other topics as well. Statistics and machine learning departments are usually too far away from the life sciences to establish joint educational programmes. As a result, scientists doing the data analysis very often have a background in biology or chemistry, and have acquired their statistical skills by training-on-the-job. This can be an advantage, since it makes it easier to interpret results and assess the relevance of certain findings. At the same time, there is a need for easily accessible background material and opportunities for self-study: books like Hastie et al. (2001) form an invaluable source of information but can also be a somewhat daunting read for scientists without much statistical background.

This book aims to fill the gap, at least to some extent. It is important to combine the sometimes rather abstract descriptions of the statistical techniques with hands-on experience behind a computer screen. In many ways R (R Development Core Team 2010) is the ideal software platform to achieve this—it is extremely powerful, the many add-on packages provide a huge range of functionalities in different areas, and it is freely accessible. As in the other books in this series, the examples can be followed step-by-step by typing or cutting-and-pasting the code, and it is easy to plug in one's own data. To date, there is only one other book specifically focused on the use of R in a similar field of science: "Introduction to Multivariate Statistical Analysis in Chemometrics" Varmuza and Filzmoser (2009) which to some extent complements the current volume, in particular in its treatment of robust statistics.

Here, the concepts behind the most important data analysis techniques will be explained using a minimum of mathematics, but in such a way that the book still can be used as a student's text. Its structure more or less follows the steps made in a "classical" data analysis, starting with the *data pretreatment* in Part I. This step is hugely important, yet is often treated only cursorily. An unfortunate choice here can destroy any hope of achieving good results: background knowledge of the system under study as well as the nature of the measurements should be used in making decisions. This is where science meets art: there are no clear-cut rules, and only by experience we will learn what the best solution is.

The next phase, subject of Part II, consists of *exploratory analysis*. What structure is visible? Are there any outliers? Which samples are very similar, which are different? Which variables are correlated? Questions like these are most easily assessed by eye—the human capacity for pattern recognition in two dimensions is far superior to any statistical method. The methods at this stage all feature strong visualization capabilities. Usually, they are model-free; no model is fitted, and the assumptions about the data are kept to a minimum.

Once we are at the *modelling* phase, described in Part III, we very often do make assumptions: some models work optimally with normally distributed data, for example. The purpose of modelling can be twofold. The first is prediction. Given a set of analytical data, we want to be able to predict properties of the samples that cannot be measured easily. An example is the assessment of whether a specific treatment will be useful for a patient with particular characteristics. Such an application is known as *classification*—one is interested in modelling class membership (will or will not respond). The other major field is *regression*, where the aim is to model continuous real variables (blood pressure, protein content, ...). Such predictive models can mean a big improvement in quality of life, and save large amounts of money. The prediction error is usually taken as a quality measure: a model that is able to predict with high accuracy must have captured some real information about the system under study. Unfortunately, in most cases no analytical expressions can be derived for prediction accuracy, and other ways of estimating prediction accuracy are required in a process called *validation*. A popular example is crossvalidation.

The second aim of statistical modelling is *interpretation*, one of the topics in Part IV. Who cares if the model is able to tell me that this is a Golden Delicious apple rather than a Granny Smith? The label in the supermarket already told me so; but the question of course is why they taste different, feel different and look different. Fitting a predictive model in such a case may still be informative: when we are able to find out why the model makes a particular prediction, we may be able to learn something about the underlying physical, chemical or biological processes. If we know that a particular gene is associated with the process that we are studying, and both this gene and another one show up as important variables in our statistical model, then we may deduce that the second gene is also involved. This may lead to several new hypotheses that should be tested in the lab. Obviously, when a model has little or no predictive ability it does not make too much sense to try and extract this type of information. The variables identified to be worth further study in many cases are indicated by the term *biomarkers*, according to Wikipedia "measurable indicators of some biological state or condition". Of course, with high-dimensional data sets it is very well possible that no biomarkers can be identified, even though predictive models can be fitted—much as we would like the world to be a simple place, it usually is not. In areas like Machine Learning one often uses models that do not allow much interpretation and consequently focus has shifted almost completely to prediction.

Our knowledge of the system can also serve as a tool to assess the quality of our model. A model that fits the data and seems to be able to predict well is not going to be very popular when its parameters contradict what we know about the underlying process. Often, prior knowledge is available (we expect a peak at a certain position; we know that model coefficients should not be negative; this coefficient should be larger than the other), and we can use that knowledge to assess the relevance of the fitted model. Alternatively, we can constrain the model in the training phase to take prior knowledge into account, which is often done with constraints. In other cases, the model is hard to interpret because of the sheer number of coefficients that have been fitted, and graphical summaries may fail to show what variables contribute in what way. In such cases, *variable selection* can come to the rescue: by discarding the majority of the variables, hopefully without compromising the model quality, one can often improve predictions *and* make the model much more easy to interpret. Unfortunately, variable selection is an NP-complete problem (which in practice means that even for moderate-sized systems it may be impossible to assess all possible solutions) and one never can be sure that the optimal solution has been found. But then again, any improvement over the original, full, model is a bonus.

For each of the stages in this "classical" data analysis pipeline, a plethora of methods is available. It can be hard to assess which techniques should be considered in a particular problem, and perhaps even more importantly, which should not. The view taken here is that the simplest possibilities should be considered first; only when the results are unsatisfactory, one should turn to more complex solutions. Of course, this is only a very crude first approach, and experienced scientists will have devised many shortcuts and alternatives that work better for their types of data. In this book, I have been forced to make choices. It is impossible to treat all methods, or even a large subset, in detail. Therefore the focus is on an ensemble of methods that will give the reader a broad range of possibilities, with enough background information to acquaint oneself with other methods, not mentioned in this book, if needed. In some cases, methods deserve a mention because of the popularity within the bioinformatics or chemometrics communities. Such methods, together with some typical applications, are treated in the final part of the book.

Given the huge number of packages available on CRAN and the speed with which new ones appear, it is impossible to mention all that are relevant to the material in this book. Where possible, I have limited myself to the recommended packages, and those coming with a default R installation. Of course, alternative, perhaps even much simpler, solutions may be available in the packages that this book does not consider. It pays to periodically scan the CRAN and Bioconductor repositories, or, e.g., check the Task Views that provide an overview of all packages available in certain areas—there is one on Physics and Chemistry, too.

Part I
Preliminaries

Chapter 2
Data

In this chapter some data sets are presented that will be used throughout the book. In a couple of places (in particular in Chap. 11) other sets will be discussed focusing on particular analysis aspects. All data sets are accessible, either through one of the packages mentioned in the text, or in the **ChemometricsWithR** package. In addition to a short description, the data will be visualized to get an idea of their form and characteristics—one cannot stress enough how important it is to eyeball the data, not only through convenient summaries but also in their raw form!

Chemical data sets nowadays are often characterized by a relatively low number of samples and a large number of variables, a result of the predominant spectroscopic measuring techniques enabling the chemist to rapidly acquire a complete spectrum for one sample. Depending on the actual technique employed, the number of variables can vary from several hundreds (typical in infrared measurements) to tens of thousands (e.g., in Nuclear Magnetic Resonance, NMR). A second characteristic is the high correlation between variables: neighboring spectral variables usually convey very similar information. An example is shown in Fig. 2.1, depicting the gasoline data set. It contains near-infrared (NIR) spectra of sixty gasolines at wavelengths from 900 to 1700 nm in 2 nm intervals (Kalivas 1997), and is available in the **pls** package. Clearly, the spectra are very smooth: there is very high correlation between neighboring wavelengths. This implies that the actual dimensionality of the data is lower than the number of variables.

The plot is made using the following piece of code:

```
> data(gasoline)
> wavelengths <- seq(900, 1700, by = 2)
> matplot(wavelengths, t(gasoline$NIR), type = "l",
+         lty = 1, xlab = "Wavelength (nm)", ylab = "1/R")
```

The `matplot` function is used to plot all columns of matrix `t(gasoline$NIR)` (or, equivalently, all rows of matrix `gasoline$NIR`) against the specified wave-

© Springer-Verlag GmbH Germany, part of Springer Nature 2020
R. Wehrens, *Chemometrics with R*, Use R!,
https://doi.org/10.1007/978-3-662-62027-4_2

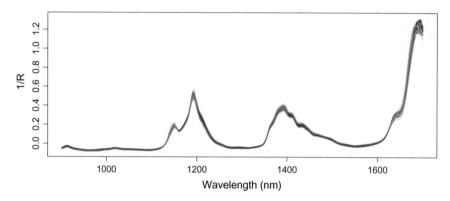

Fig. 2.1 Near-infrared spectra of sixty gasoline samples, consisting of 401 reflectance values measured at equally spaced wavelengths between 900 and 1700 nm

lengths. Clearly, all samples have very similar features—it is impossible to distinguish individual samples in the plot. NIR spectra are notoriously hard to interpret: they consist of a large number of heavily overlapping peaks which leads to more or less smooth spectra. Nevertheless, the technique has proven to be of immense value in industry: it is a rapid, non-destructive method of analysis requiring almost no sample preprocessing, and it can be used for quantitative predictions of sample properties. The data used here can be used to quantitatively assess the octane number of the gasoline samples, for instance.

In other cases, specific variables can be directly related to absolute or relative concentrations. An example in which is the case for most variables is the wine data set from the **kohonen** package, used throughout the book. It is a set consisting of 177 wine samples, with thirteen measured variables (Forina et al. 1986):

```
> data(wines)
> colnames(wines)
 [1] "alcohol"          "malic acid"        "ash"
 [4] "ash alkalinity"   "magnesium"         "tot. phenols"
 [7] "flavonoids"       "non-flav. phenols" "proanth"
[10] "col. int."        "col. hue"          "OD ratio"
[13] "proline"
```

Variables are reported in different units. All variables apart from `"col. int."`, `"col. hue"` and `"OD ratio"` are concentrations. The meaning of the variables color intensity and color hue is obvious; the OD ratio is the ratio between the absorbance at wavelengths 280 and 315 nm. All wines are from the Piedmont region in Italy. Three different classes of wines are present: Barolo, Grignolino and Barberas. Barolo wine is made from Nebbiolo grapes; the other two wines have the name of the grapes from which they are made. Production areas are partly overlapping (Forina et al. 1986).

```
> table(vintages)
vintages
   Barbera      Barolo Grignolino
        48          58          71
```

The obvious aim in the analysis of such a data set is to see whether there is any structure that can be related to the three cultivars. Possible questions are: "which varieties are most similar?", "which variables are indicative of the variety?", "can we discern subclasses within varieties?", etcetera.

A quick overview of the first few variables can be obtained with a so-called pairs plot:

```
> wine.classes <- as.integer(vintages)
> pairs(wines[, 1:3], pch = wine.classes, col = wine.classes)
```

This leads to the plot shown in Fig. 2.2. It is clear that the three classes can be separated quite easily—consider the plot of alcohol against malic acid, for example.

A further data set comes from the field of mass-spectrometry-based proteomics.[1] Figure 2.3, showing the first mass spectrum (a healthy control sample) is generated by:

```
> data(Prostate2000Raw)
> plot(Prostate2000Raw$mz, Prostate2000Raw$intensity[, 1],
+      type = "h", main = "Prostate data",
+      xlab = bquote(italic(.("m/z"))~.("(Da)")),
+      ylab = "Intensity")
```

Each peak in the chromatogram corresponds to the elution of a compound, or in more complex cases, a number of overlapping compounds. In a process called peak picking (see next chapter) these peaks can be easily quantified, usually by measuring peak area, but sometimes also by peak height. Since the number of peaks usually is orders of magnitude smaller than the number of variables in the original data, summarising the chromatograms with a peak table containing position and intensity information can lead to significant data compression. Mass spectra, containing intensities for different mass-to-charge ratios (indicated by m/z), can be recorded at a very high resolution. To enable statistical analysis, m/z values are typically *binned* (or "bucketed"). Even then, thousands of variables are no exception.

The data set contains 327 samples from three groups: patients with prostate cancer, benign prostatic hyperplasia, and normal controls (Adam et al. 2002; Qu et al. 2002). All samples have been measured in duplicate:

---

[1] Originally from the R package **msProstate**, which is no longer available.

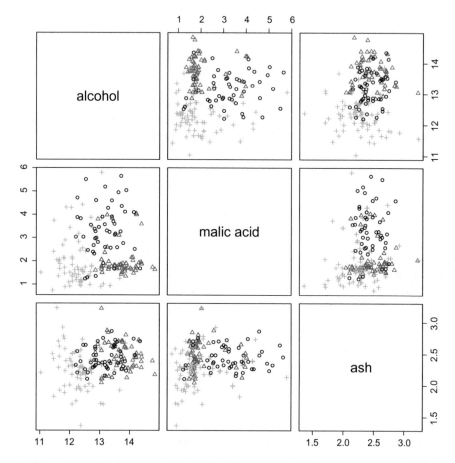

Fig. 2.2 A pairs plot of the first three variables of the wine data. The three vintages are indicated with different colors and plotting symbols: Barbera wines are indicated with black circles, Barolos with red triangles and Grignolinos with green plusses

```
> table(Prostate2000Raw$type)

  bph control    pca
  156     162    336
```

The data have already been preprocessed (binned, baseline-corrected, normalized—see Chap. 3); m/z values range from 200 to 2000 Dalton.

Such data can serve as diagnostic tools to distinguish between healthy and diseased tissue, or to differentiate between several disease states. The number of samples is

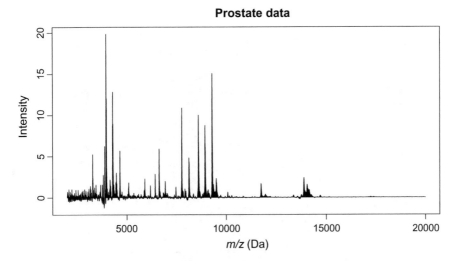

Fig. 2.3 The first mass spectrum in the prostate MS data set

almost always very low—for rare diseases, patients are scarce, and stratification to obtain relatively homogeneous groups (age, sex, smoking habits, ...) usually does the rest; and in cases where the measurement is unpleasant or dangerous it may be difficult or even unethical to get data from healthy controls. On the other hand, the number of variables per sample is often huge. This puts severe restrictions on the kind of analysis that can be performed and makes thorough validation even more important.

The final data set in this chapter also comes from proteomics and is measured with LC-MS, the combination of liquid chromatography and mass spectrometry. The chromatography step serves to separate the components of a mixture on the basis of properties like polarity, size, or affinity. At specific time points a mass spectrum is recorded, containing the counts of particles with specific m/z values. Measuring several samples therefore leads to a data cube of dimensions ntime, nmz, and nsample; the number of time points is typically in the order or thousands, whereas the number of samples rarely exceeds one hundred. Package **ptw** provides a data set, lcms, containing data on three tryptic digests of E. coli proteins (Bloemberg et al. 2010).

Figure 2.4 shows a top view of the first sample. The projection to the top of the figure, effectively summing over all m/z values, leads to the "Total Ion Current" (TIC) chromatogram. Similarly, if the chromatographic dimension would be absent, the mass spectrum of the whole sample would be very close to the projection on the right (a "direct infusion" spectrum). The whole data set consists of three of such

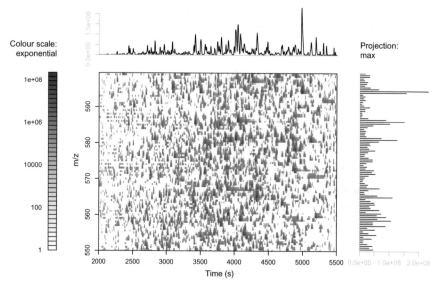

Fig. 2.4 Top view of the first sample in data set `lcms`. The TIC chromatogram is shown on the top, and the direct infusion mass spectrum on the right

planes, leading to a data cube of size $100 \times 2000 \times 3$. Similar data sets are seen in the field of metabolomics, where the chemical entities that are sampled are not peptides (small fragments of proteins) as in proteomics, but small chemical molecules called metabolites. Because of the high dimensionality and general complexity of such data sets, chemometric methods have caught on very well in the -omics sciences.

Chapter 3
Preprocessing

Textbook examples typically use clean, perfect data, allowing the techniques of interest to be explained and illustrated. However, in real life data are messy, noisy, incomplete, downright faulty, or a combination of these. The first step in any data analysis often consists of preprocessing to assess and possibly improve data quality. This step may actually take more time than the analysis itself, and more often than not the process consists of an iterative procedure where data preprocessing steps are alternated with data analysis steps.

Some problems can immediately be recognized, such as measurement noise, spikes, non-detects, and unrealistic values. In these cases, taking appropriate action is rarely a problem. More difficult are the cases where it is not obvious which characteristics of the data contain information, and which do not. There are many examples where chance correlations lead to statistical models that are perfectly able to describe the training data (the data used to set up the model in the first place) but have no predictive abilities whatsoever.

This chapter will focus on standard preprocessing techniques used in the natural sciences and the life sciences. Data are typically spectra or chromatograms, and topics include noise reduction, baseline removal, peak alignment, peak picking, and scaling. Only the basic general techniques are mentioned here; some more specific ways to improve the quality of the data will be treated in later chapters. Examples include Orthogonal Partial Least Squares for removing uncorrelated variation (Sect. 11.4) and variable selection (Chap. 10).

3.1 Dealing with Noise

Physico-chemical data always contain noise, where the term "noise" is usually reserved for small, fast, random fluctuations of the response. The first aim of any scientific experiment is to generate data of the highest quality, and much effort is

© Springer-Verlag GmbH Germany, part of Springer Nature 2020
R. Wehrens, *Chemometrics with R*, Use R!,
https://doi.org/10.1007/978-3-662-62027-4_3

usually put into decreasing noise levels. The simplest experimental way is to perform n repeated measurements, and average the individual spectra, leading to a noise reduction with a factor $\sqrt{n}$. In NMR spectroscopy, for example, a relatively insensitive analytical method, signal averaging is routine practice, where one has to strike a balance between measurement time and data quality.

As an example, we consider the prostate data, where each sample has been measured in duplicate. The replicate measurements of the prostate data cover consecutive rows in the data matrix. Averaging can be done using the following steps:

```
> prostate.array <- array(t(Prostate2000Raw$intensity),
+                               c(2, 327, 10523))
> prostate <- apply(prostate.array, c(2, 3), mean)
```

The idea is to convert the matrix into an array where the first dimension contains the two replicates for every sample—each element in the first dimension thus contains one complete set of measurements. The final data matrix is obtained by averaging the replicates. In this code snippet the outcome of `apply` is a matrix having dimensions equal to the second and third dimensions of the input array. The first dimension is averaged out. As we will see later, `apply` is extremely handy in many situations. There is, however, a much faster possibility using `rowsum` leading to the same result:

```
> prostate <- rowsum(t(Prostate2000Raw$intensity),
+                          group = rep(1:327, each = 2),
+                          reorder = FALSE) / 2
> dim(prostate)
[1]    327 10523
```

This function sums all the rows for which the grouping variable (the second argument) is equal. Since there are two replicates for every sample, the result is divided by two to get the average values, and is stored in variable `prostate`. For this new variable we should also keep track of the corresponding class labels:

```
> prostate.type <- Prostate2000Raw$type[seq(1, 654, by = 2)]
```

The result of the signal averaging is visualized in Fig. 3.1, using the function `matplot` to show the original data, and afterwards adding the averaged data to make sure they are plotted on top. For clarity, we are limiting ourselves to the first 250 m/z values:

```
> matplot(Prostate2000Raw$mz[1:250],
+         Prostate2000Raw$intensity[1:250, 1:2],
+         type = "l", col = "pink", lty = 1,
+         xlab = bquote(italic(.("m/z"))~.("(Da)")),
+         ylab = "response")
> lines(Prostate2000Raw$mz[1:250], prostate[1, 1:250], lwd = 2)
```

Fig. 3.1 The first averaged mass spectrum in the prostate data set; only the first 250 m/z values are shown. The original data are shown in pink

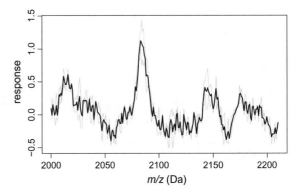

Also in the averaged data the noise is appreciable; reducing the noise level while taking care not to destroy the data structure would make subsequent analysis much easier.

The simplest approach is to apply a *running mean*, i.e., to replace every single value by the average of the k points around it. The value of k, the so-called window size, needs to be optimized; large values lead to a high degree of smoothing, but also to peak distortion, and low values of k can only make small changes to the signal. Very often k is chosen on the basis of visual inspection, either of the smoothed signal itself or of the residuals. Running means can be easily calculated using the function embed, providing a matrix containing successive chunks of the original data vector as rows; using the function rowMeans one then can obtain the desired running means.

```
> rmeans <- rowMeans(embed(prostate[1, 1:250], 5))
> plot(Prostate2000Raw$mz[1:250], prostate[1, 1:250],
+      type = "l", xlab = "m/z", ylab = "response",
+      main = "running means", col = 2)
> lines(Prostate2000Raw$mz[3:248], rmeans, type = "l", lwd = 2)
```

As can be seen in the left plot in Fig. 3.2, the smoothing effectively reduces the noise level. Note that the points at the extremes need to be treated separately in this implementation. The price to be paid is that peak heights are decreased, and especially with larger spans one will see appreciable peak broadening. These effects can sometimes be countered by using running medians instead of running means. The function runmed, part of the **stats** package, is available for this:

```
> plot(Prostate2000Raw$mz[1:250], prostate[1, 1:250],
+      type = "l", xlab = "m/z", ylab = "response",
+      main = "running median", col = 2)
> lines(Prostate2000Raw$mz[1:250],
+       runmed(prostate[1, 1:250], k = 5), type = "l", lwd = 2)
```

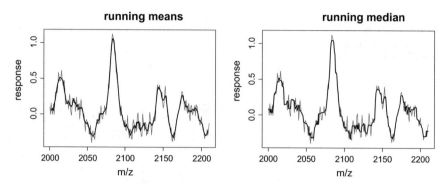

Fig. 3.2 Smoothing (black lines) of the averaged mass spectrum of Fig. 3.1 (in red): a running mean (left plot) and a running median (right plot), both with a window size of five

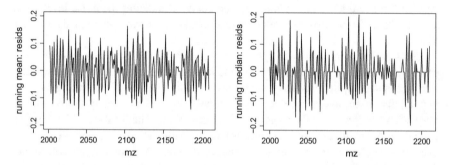

Fig. 3.3 Residuals of smoothing the first spectrum of the prostate data with a running mean (left plot) or running median (right)

The result is shown in the right plot in Fig. 3.2. Its appearance is less smooth than the running mean with the same window size. Note that the function `runmed` does return a vector with the same length as the input—the points at the extremes are unchanged. The plots of the residuals in Fig. 3.3 show that both smoothing techniques do quite a good job in removing high-frequency noise components without distorting the signal too much: there is not much structure left in the residual plots.

Many other smoothing functions are available—only a few will be mentioned here briefly. In signal processing, Savitsky–Golay filters are a popular choice (Savitsky and Golay 1964); every point is replaced by a smoothed estimate obtained from a local polynomial regression. An added advantage is that derivatives can simultaneously be calculated (see below). In statistics, robust versions of locally weighted regression (Cleveland 1979) are often used; `loess` and its predecessor `lowess` are available in R as simple-to-use implementations. The fact that the fluctuations in the noise usually are much faster than the data has led to a whole class of frequency-based smoothing methods, of which wavelets (Nason 2008) are perhaps the most popular ones. The idea is to set the coefficients for the high-frequency components to zero, which should leave only the signal component.

Fig. 3.4 Binned version of the mass-spectrometry data from Fig. 3.1. Five data points constitute one bin

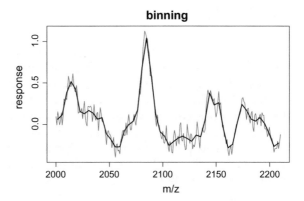

A special kind of smoothing is provided by *binning*, also called bucketing, which not only averages consecutive values but also decreases the number of variables. To replace five data points with their average, one can use:

```
> mznew <- colMeans(matrix(Prostate2000Raw$mz[1:250], nrow = 5))
> xnew <- colMeans(matrix(prostate[1, 1:250], nrow = 5))
> plot(Prostate2000Raw$mz[1:250], prostate[1, 1:250],
+      type = "l", xlab = "m/z", ylab = "response",
+      main = "binning", col = 2)
> lines(mznew, xnew, lwd = 2)
```

We have seen the idea before: in this case we fill a matrix with five rows column-wise with the data, and then average over the rows. This leads to the plot in Fig. 3.4.

Obviously, the binned representation gives a cruder description of the data, but still is able to follow the main features. Again, determining the optimal bin size is a matter of trial and error. Binning has several major advantages over running means and medians. First, it can be applied when data are not equidistant and even when the data are given as positions and intensities of features, as is often the case with mass-spectrometric data. Second, the effect of peak shifts (see below) is decreased: even when a peak is slightly shifted, it will probably be still within the same bin. And finally, more often than not the variable-to-object ratio is extremely large in data sets from the life sciences. Summarizing the information in fewer variables in many cases makes the subsequent statistical modelling more easy.

Although smoothing leads to data that are much better looking, one should also be aware of the dangers. Too much smoothing will remove features, and even when applied prudently, the noise characteristics of the data will be different. This may significantly affect statistical modelling.

3.2 Baseline Removal

In some forms of spectroscopy one can encounter a baseline, or "background signal" that is far away from the zero level. Since this influences measures like peak height and peak area, it is of utmost importance to correct for such phenomena.

Infrared spectroscopy, for instance, can lead to scatter effects—the surface of the sample influences the measurement. As a result, one often observes spectral offsets: two spectra of the same material may show a constant difference over the whole wavelength range. This may be easily removed by taking first derivatives (i.e., looking at the *differences* between intensities at sequential wavelengths, rather than the intensities themselves). Take a look at the gasoline data:

```
> nir.diff <- t(apply(gasoline$NIR, 1, diff))
> matplot(wavelengths[-1] + 1, t(nir.diff),
+          xlab = "Wavelength (nm)", ylab = "1/R (1st deriv.)",
+          type = "n")
> abline(h = 0, col = "gray")
> matlines(wavelengths[-1] + 1, t(nir.diff), lty = 1)
```

Note that the number of variables decreases by one. The result is shown in Fig. 3.5. Comparison with the original data (Fig. 2.1) shows more detailed structure; the price is an increase in noise. A better way to obtain first-derivative spectra is given by the Savitsky–Golay filter (here using the `sgolayfilt` function from the **signal** package), which is not only a smoother but can also be used to calculate derivatives:

```
> nir.deriv <- apply(gasoline$NIR, 1, sgolayfilt, m = 1)
```

In this particular case, the differences between the two methods are very small. Also second derivatives are used in practice—the need to control noise levels is even bigger in that case.

Another way to remove scatter effects in infrared spectroscopy is Multiplicative Scatter Correction (MSC, Geladi et al. 1985; Næs et al. 1990). One effectively models

Fig. 3.5 First-derivative representation of the gasoline NIR data

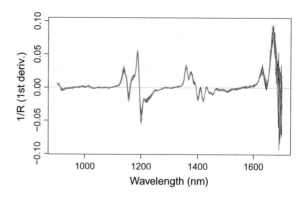

the signal of a query spectrum as a linear function of the reference spectrum:

$$y_q = a + by_r$$

An obvious reference spectrum may not be available, and then often a mean spectrum is used. This is also the approach in the msc function of the **pls** package:

```
> nir.msc <- msc(gasoline$NIR)
```

For the gasoline data, the effects of MSC are quite small.

In more difficult cases, a non-constant baseline drift can be observed. First derivatives are not enough to counter such effects, and one has to resort to techniques that actually estimate the shape of the baseline. The exact function is usually not important—the baseline will be subtracted and that is it. To illustrate this point, consider the first chromatogram in the prostate data. One very simple solution is to connect local minima, obtained from, e.g., 200-point sections:

```
> x <- Prostate2000Raw$intensity[1:4000, 1]
> mz <- Prostate2000Raw$mz[1:4000]
> lsection <- 200
> xmat <- matrix(x, nrow = lsection)
> ymin <- apply(xmat, 2, min)
> plot(mz, x, type = "l", col = "darkgray", ylim = c(-1.2, 5),
+       xlab = "m/z", ylab = "I")
> lines(mz, rep(ymin, each = lsection))
```

We have used the by now familiar trick to convert a vector to a matrix and calculate minimal values for every column to obtain the intensity levels of the horizontal line segments. The result is shown as the stair-like series of line segments in the left plot of Fig. 3.6. Obviously, a more smooth baseline estimate would be better. One function that can be used is loess, fitting local polynomials through the minimal values:

```
> bsln.loess <- loess(ymin ~ mz[seq(101, 4000, by = 200)])
> lines(mz, predict(bsln.loess, mz), lwd = 2, col = "blue")
```

This leads to the smooth blue line in the left plot in Fig. 3.6. Note that the line is not influenced much by local variations, such as the large dip near 4000 m/z. For a baseline estimate that is probably all right—by tweaking the smoother settings one can adjust the result to the characteristics of the data to obtain a solution that makes most sense. A similar effect could be obtained for the local-minimum approach in the same figure by changing the length of the horizontal sections: a smaller length would lead to more detail, and a larger length to a more global baseline estimate.

Another alternative is to use asymmetric least squares, where deviations above the fitted curve are not taken into account (or only with a very small weight). This is implemented in function baseline.corr in the **ptw** package, which returns

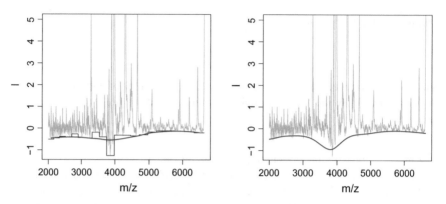

Fig. 3.6 Simple baseline correction for the first mass spectrum in the `prostate` data: in the left plot the baseline is estimated by a series of twenty local minima, the connected horizontal segments. The blue line indicates the loess smooth (using default settings) of these minima. Right plot: asymmetric least squares estimate of the baseline

a baseline-corrected signal. Internally, it uses the function `asysm` to estimate the baseline:

```
> plot(mz, x, col = "darkgray", type = "l", ylim = c(-1.2, 5),
+      xlab = "m/z", ylab = "I")
> lines(mz, asysm(x), lwd = 2, col = "blue")
```

The result is shown in the right plot of Fig. 3.6. Again, the parameters of the `asysm` function may be tweaked to get optimal results.

Obviously, many more techniques can be used to estimate and remove baselines—wavelets are a popular approach (Nason 2008).

3.3 Aligning Peaks—Warping

Many analytical data suffer from small shifts in peak positions. In NMR spectroscopy, for example, the position of peaks may be influenced by the pH. What complicates matters is that in NMR, these shifts are by no means uniform over the data; rather, only very few peaks shift whereas the majority will remain at their original locations. The peaks may even move in different directions. In mass spectrometry, the shift is more uniform over the m/z axis and is more easy to account for—if one aims to analyse the data in matrix form, binning is required, and in many cases a suitable choice of bins will already remove most if not all of the effects of shifts. Moreover, peak shifts are usually small, and may be easily corrected for by the use of standards.

The biggest shifts, however, are encountered in chromatographic applications, especially in liquid chromatography. Two different chromatographic columns almost never give identical elution profiles, up to the extent that peaks may even swap posi-

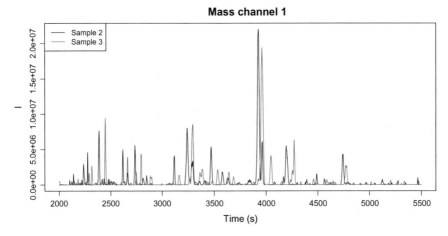

Fig. 3.7 Comparison of two chromatograms of the `lcms` data set. Clearly, corresponding features are not in the same positions

tions. The situation is worse than in gas chromatography, since retention mechanisms are more complex in the liquid phase than in the gas phase. In all forms of column chromatography, column age is an important factor: a column that has been used for some time almost certainly will show different chromatograms than when freshly installed.

Peak shifts pose significant problems in modelling. In Fig. 3.7 the chromatograms of the first mass channel in two of the samples of the `lcms` data (from the **ptw** package) are shown:

```
> plot(time, lcms[1, , 2], type = "l",
+      xlab = "Time (s)", ylab = "I", main = "Mass channel 1")
> lines(time, lcms[1, , 3], type = "l", col = 2)
> legend("topleft", legend = paste("Sample", 2:3),
+        lty = 1, col = 1:2)
```

Clearly, both chromatograms contain the same features, although at different locations—the shift is, equally clearly, not constant over the whole range. Comparing these chromatograms with a distance-based similarity function based on Euclidean distance will lead to the conclusion that there are huge differences, whereas the chromatograms in reality are very similar.

Correction of such shifts is known as "time warping", one of the more catchy names in data analysis. The technique originates in speech processing (Sakoe and Chiba 1978; Rabiner et al. 1978), and nowadays many forms exist. The most popular in the literature for natural sciences and life sciences are Dynamic Time Warping (DTW, Wang and Isenhour 1987), Correlation-Optimized Warping (COW, Nielsen et al. 1998) and Parametric Time Warping (PTW, Eilers 2004). Optimization methods are used to align two signals, often using the squared differences between the two as a criterion; this is the case for DTW and the original version of PTW (Eilers 2004). As an alternative, the R package **ptw** (Bloemberg et al. 2010; Wehrens et al. 2015a) provides the so-called *weighted cross correlation* (WCC, de Gelder et al. 2001) criterion to assess the similarity of two shifted patterns. In this context the WCC is used as a distance measure so that a value of zero indicates perfect alignment (Bloemberg et al. 2010). COW maximizes the correlation between patterns, where the signals are cut into several segments which are treated separately. Another package, **VPdtw** (Clifford et al. 2009; Clifford and Stone 2012), implementing a penalized form of dynamic time warping is available from http://www.github.com/david-clifford.

3.3.1 Parametric Time Warping

In PTW, one approximates the time axis of the reference signal by applying a polynomial transformation of the time axis of the sample (Eilers 2004):

$$\hat{S}(t_k) = S(w(t_k)) \tag{3.1}$$

where $\hat{S}(t_k)$ is the value of the warped signal at time point t_k, where k is an index. The warping function, w, is given by:

$$w(t) = \sum_{j=0}^{J} a_j t^j \tag{3.2}$$

with J the maximal order of the polynomial. In general, only low-order polynomials are used. Since neighboring points on the time axis will be warped with almost the same amount, peak shape distortions are limited. Thus, the method finds the set of coefficients $a_0, \ldots, a_J$ that minimizes the difference between the sample S and reference R, using whatever difference measure is desired. Especially for higher-degree warpings there is a real possibility that the optimization ends in a local optimum, and it is usually a good idea to use several different starting values.

This procedure is very suitable for modelling gradual changes, such as the slow deterioration of chromatographic columns, so that measurements taken days or weeks apart can still be made comparable. For situations where a few individual peak shifts have to be corrected (e.g., pH-dependent shifting of patterns in NMR spectra), the technique is less ideal (Giskeødegård et al. 2010).

The original implementation of ptw (corresponding to function argument mode = "backward") predicts, for position i, which point j in the signal will end up at position i. This is somewhat counterintuitive, and in later versions (from version 1.9.1 onwards) the default mode is "forward", basically predicting the position of point i after warping. The interpretation of the coefficients in the two modes is the same, just with reversed signs.

We will illustrate the use of PTW by aligning the first mass chromatograms of Fig. 3.7. Sample number 2 will be used as a reference, and the third sample is warped so that the peak positions show maximal overlap:

```
> sref <- lcms[1, , 2]
> ssamp <- lcms[1, , 3]
> lcms.warp <- ptw(sref, ssamp, init.coef = c(0, 1, 0))
> summary(lcms.warp)
PTW object: individual alignment of 1 sample on 1 reference.

Warping coefficients:
        [,1]   [,2]          [,3]
[1,] -43.047 1.0272 -5.8906e-06

Warping criterion: WCC
Warping mode: forward
Value: 0.087797
```

Using the default quadratic warping function with initial values init.coef = c(0, 1, 0), corresponding to the unit warp (no shift, unit stretch, no quadratic warping), we arrive at a warping where the sample is shifted approximately 43 points to the left, is compressed around 3%, and experiences also a small quadratic warping. The result is an agreement of 0.088, according to the default WCC criterion. A visual check, shown in Fig. 3.8, confirms that the peak alignment is much improved:

```
> plot(time, sref, type = "l", lwd = 2, col = 2,
+      xlim = time[c(600, 1300)], xlab = "Time (s)", ylab = "I")
> lines(time, ssamp + 1e6, lty = 2)
> lines(time, lcms.warp$warped.sample + 2e6, col = 4)
> legend("topleft", lty = c(1, 2, 1), col = c(2, 1, 4),
+        legend = c("Reference", "Sample", "Warped sample"),
+        lwd = c(2, 1, 1))
```

To show the individual traces more clearly, a small vertical offset has been applied to both the unwarped and warped sample. Obviously, the biggest gains can be made at the largest peaks, and in the warped sample the features around 3250 and 3900 s are aligned really well. Nevertheless, some other features, such as the peaks between 3450 and 3650 s, and the peaks at 3950 and 4200 s still show shifts. This simple quadratic warping function apparently is not flexible enough to iron out these differences. Note that squared-difference-based warping in this case leads to very similar results:

```
> lcms.warpRMS <- ptw(sref, ssamp, optim.crit = "RMS")
> lcms.warpRMS$warp.coef
        [,1]   [,2]          [,3]
[1,] -39.011 1.0169 -1.4243e-06
```

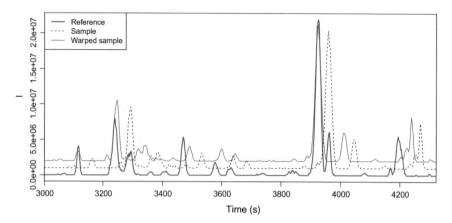

Fig. 3.8 PTW of the data shown in Fig. 3.7, using a quadratic warping function. For easier interpretation, small vertical offsets have been added to the sample and warped sample spectra and only
a part of the time axis is shown

More complex warping functions, fitting polynomials of degrees three to five, can
be tried:

```
> lcms.warp2 <- ptw(sref, ssamp, init = c(0, 1, 0, 0))
> lcms.warp3 <- ptw(sref, ssamp, init = c(0, 1, 0, 0, 0))
> lcms.warp4 <- ptw(sref, ssamp, init = c(0, 1, 0, 0, 0, 0))
```

To visualize these warping functions, we first gather all warpings in one list, and
obtain the qualities of each element using a close relative of the `apply` function,
`sapply`:

```
> allwarps <- list(lcms.warp, lcms.warp2, lcms.warp3, lcms.warp4)
> wccs <- sapply(allwarps, function(x) x$crit.value)
```

Where `apply` operates on rows or columns of a matrix, `sapply` performs actions
on list elements, and returns the result in a simple way, in this case a matrix.

Because we are interested in the deviations from the identity warp (i.e., no change),
we subtract that from the warping functions:

```
> allwarp.funs <- sapply(allwarps, function(x) x$warp.fun)
> warpings <- allwarp.funs - 1:length(sref)
```

Finally, we can plot the columns of the resulting matrix:

```
> matplot(time, warpings, type = "l", lty = rep(c(1, 2), 2),
+         col = 1:4, ylim = c(min(warpings), 0.5))
> abline(h = 0, col = "gray", lty = 2)
> legend("bottom", lty = rep(c(1, 2), 2), col = 1:4, bty = "n",
+        legend = paste("Degree", 2:5, " - WCC =", round(wccs, 3)))
```

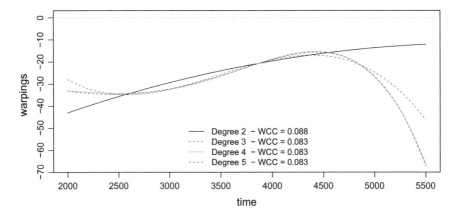

Fig. 3.9 PTW warping functions of different degrees

This leads to Fig. 3.9. The horizontal line at zero indicates the identity warp. The higher-degree warping functions lead to more or less the same values for the WCC criterion; the functions for degrees four and five are almost identical. Note that all these lead to much larger warpings in the right part of the chromatogram than the second-degree warping function, a sign that these models are trying to correct also the differences at later elution times. Note that individual warping functions can be shown very easily by a command like

```
> plot(lcms.warp, "function")
```

One very important characteristic of LC-MS data is that it contains multiple m/z traces. If the main reason for differences in retention time is the state of the column, all traces should be warped with the same warping function. When using multiple traces, one should have less trouble identifying the optimal warping since ambiguities, that may exist in single chromatograms, are resolved easily (Bloemberg et al. 2010). If all traces should be warped with the same coefficients, one could speed up the procedure by performing the warping on a subset of samples—bad-quality traces can have a bad influence on warping estimates, so it makes sense to use a high-quality subset. Several criteria exist to choose such a subset, e.g., choosing traces with high average intensity. A popular criterion that is slightly more sophisticated is Windig's Component Detection Algorithm (CODA, Windig et al. 1996), one of the choices in **ptw**'s select.traces function. CODA basically selects high-variance traces after smoothing and a unit-length scaling step—the variance of a trace with one high-intensity feature will always be larger than that of a trace with the same length but many low-intensity features. The following example chooses the ten traces which show the highest values for the CODA criterion, and uses these for constructing one global warping function:

```
> srefM <- lcms[, , 2]
> ssampM <- lcms[, , 3]
> traces <- select.traces(srefM, criterion = "var")
> lcms.warpglobal <-
+    ptw(srefM, ssampM, warp.type = "global",
+         selected.traces = traces$trace.nrs[1:10])


> summary(lcms.warpglobal)
PTW object: global alignment of 10 samples on 10 references.

Warping coefficients:
          [,1]    [,2]           [,3]
[1,] -95.514 1.1292 -5.8147e-05

Warping criterion: WCC
Warping mode: forward
Value: 0.13845
```

In other cases, one would like to have individual warpings for individual traces; this is the default mode for applying the `ptw` function (`warp.type = "individual"`).

Finally, one can also mix global and individual alignments. Here we will compare an eight-degree individual warping with a global warping of degree two, followed by a four-degree individual warping, also leading to a net warping degree of eight (Bloemberg et al. 2010). Because repeated application of ptw can lead to information loss at the extremes of the data points (compression upon compression) we first add columns of zeros to both sides of the data matrices:

```
> npad <- 500
> srefM2 <- padzeros(lcms[, , 2], npad, "both")
> ssampM2 <- padzeros(lcms[, , 3], npad, "both")
> sample2.indiv.warp <-
+    ptw(srefM2, ssampM2, init.coef = c(0, 1, 0, 0, 0, 0, 0, 0, 0))
> sample2.global.warp <-
+    ptw(srefM2, ssampM2, init.coef = c(0, 1, 0),
+        warp.type = "global")
> sample2.final.warp <-
+    ptw(srefM2, ssampM2,
+        init.coef = c(sample2.global.warp$warp.coef, 0, 0))
```

The individual warping is initialized using estimates for the lower degree coefficients found in the global warping. We evaluate the results by looking at the total ion currents, given by the column sums of the warped samples:

PTW (indiv.)

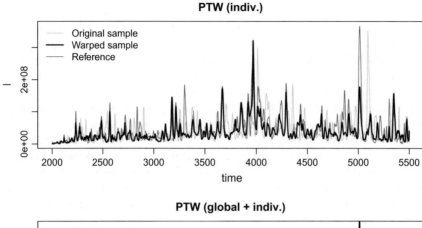

PTW (global + indiv.)

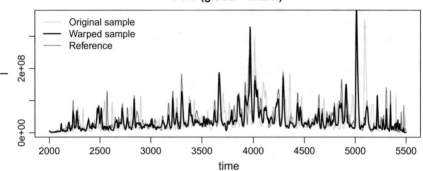

Fig. 3.10 Comparison of individual and global parametric time warping for samples two and three of the `lcms` data: the total ion current (TIC) of the aligned samples is shown. In the top panel, all mass traces have been warped individually using a warping function of degree eight—the bottom panel shows a two-stage warping using a global warping of degree two, followed by an individual warping of degree four

```
> orig.ind <- (npad + 1):(npad + 2000)
> plot(time, colSums(srefM2)[orig.ind], col = 2,
+      type = "l", main = "PTW (indiv.)", ylab = "I")
> lines(time, colSums(ssampM2)[orig.ind], col = "gray")
> lines(time,
+      colSums(sample2.indiv.warp$warped.sample)[orig.ind], lwd = 2)
> legend("topleft", bty = "n", lty = 1, lwd = c(1, 2, 1),
+      legend = c("Original sample", "Warped sample", "Reference"),
+      col = c("gray", "black", "red"))
```

This gives the top panel in Fig. 3.10, showing the result of the individual eight-degree warping. The bottom plot is produced with similar code, using `sample2.final.warp` instead of `sample2.indiv.warp`. Clearly, both alignments lead to an overall result that is very good.

Warping a study sample to a reference will only be successful if the same features, or at least a reasonable number of them, are present in both samples. However, in fields like metabolomics this is not always the case: features present in one sample may be absent from another, and standard application of warping methodology can easily lead to a situation where the wrong peaks are matched up. A similar situation can exist in comparing two distinctly different groups, e.g., treated samples and controls (Wehrens et al. 2015a). If quality control (QC) samples are available, typically the same pooled sample injected several times during the injection sequence, a good strategy is to base a warping profile on aligning these QC samples. For aligning a particular study sample, one can then use an interpolation of the warping functions used for the surrounding QCs. In Wehrens et al. (2015a), the results were very good, even though large differences in retention times were observed in the raw data due to a leakage in the HPLC system leading to increasing pressure differences during the experiment.

A final remark is that the **ptw** package also supports warpings of non-continuous data (Wehrens et al. 2015a) such as peak lists obtained by peak-picking procedures (see next paragraph). Not only is this procedure much faster than warping full profiles, it is also more robust in that baselines and other strange effects not corresponding to true features are eliminated in the peak-picking phase and no longer influence the warping.

3.3.2 Dynamic Time Warping

Dynamic Time Warping (DTW), implemented in package **dtw** (Giorgino 2009), provides a similar approach, constructing a warping function that provides a mapping from the indices in the query signal to the points in the reference signal[1]:

```
> warpfun.dtw <- dtw(ssamp, sref)
> plot(warpfun.dtw)
> abline(0, 1, col = "gray", lty = 2)
```

The result is shown in the left plot of Fig. 3.11. Here, the warping function is not restricted to be a polynomial, as in PTW.

A horizontal segment indicates that several points in the query signal are mapped to the same point in the reference signal; the axis of the query signal is compressed by elimination (or rather, averaging) of points. Similarly, vertical segments indicate a stretching of the query signal axis by the duplication of points. Note that these horizontal and vertical regions in the warping function of Fig. 3.11 may also lead to peak shape distortions.

DTW chooses the warping function that minimizes the (weighted) distance between the warped signals:

---

[1]Note that the order of sample and reference arguments is reversed compared to the ptw function.

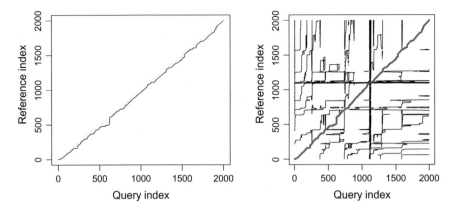

Fig. 3.11 Left plot: warping function of the data from Fig. 3.7. The identity warp is indicated with a dashed gray line. Right plot: contour lines of the cost function, and the final warping function (fat line)

$$\sum_k \{q(m(k)) - r(n(k))\}^2 w(k) / \sum_k w(k)$$

where k is the common axis to which both the query q and the reference r are mapped, $m(k)$ is the warped query signal and $n(k)$ is the warped reference. Note that in this symmetric formulation there is no difference in treatment of query and reference signals: reversing the roles would lead to the same mapping. The weights are used to remove the tendency to select the shortest warping path, but should be chosen with care. The weighting scheme in the original publication (Sakoe and Chiba 1978) is for point $k + 1$:

$$w(k + 1) = m(k + 1) - m(k) + n(k + 1) - n(k)$$

That is, if both indices advance to the next point, the weight is 2; if only one of the indices advances to the next point, the weight is 1. A part of the cumulative distance from the start of both signals is shown in the right plot of Fig. 3.11: the warping function finds the minimum through the (often very noisy) surface.

Obviously, such a procedure is very flexible, and indeed, one can define warping functions that put any two signals on top of each other, no matter how different they are. This is of course not what is desired, and usually several constraints are employed to keep the warping function from extreme distortions. One can, e.g., limit the maximal warping, or limit the size of individual warping steps. The dtw package implements these constraints and also provides the possibility to align signals of different length.

Once the warping function is calculated, we can use it to actually map the points in the second signal to positions corresponding to the first. For this, the warp function

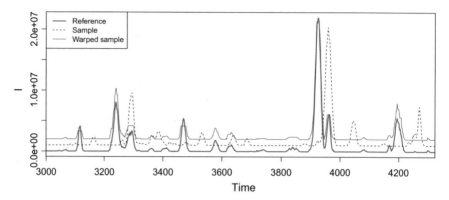

Fig. 3.12 DTW-corrected mass-spectrometry data

should be used, which internally performs a linear interpolation of the common axis
to the original axes:

```
> warped.signal <- warp(warpfun.dtw, index.reference = FALSE)
> plot(time, sref, type = "l", lwd = 2, col = 2,
+       xlim = c(time[600], time[1300]),
+       xlab = "Time", ylab = "I")
> lines(time, ssamp + 1e6, lty = 2)
> lines(time, ssamp[warped.signal] + 2e6, col = 4)
> legend("topleft", lty = c(1, 2, 1), col = c(2, 1, 4),
+        legend = c("Reference", "Sample", "Warped sample"),
+        lwd = c(2, 1, 1))
```

The warped signal can directly be compared to the reference. The result is shown
in Fig. 3.12. Immediately one can see that the warping is perfect: peaks are in exactly
the same positions. The only differences between the two signals are now found in
areas where the peaks in the reference signal are higher than in the warped signal
(e.g., m/z values at 3,100 and just below 3,300)—these peak distortions can not be
corrected for.

These data are ideally suited for DTW: individual mass traces contain not too
many, nicely separated peaks, so that it is clear what features should be aligned. The
quality of the warping becomes clear when we align all traces individually and then
compare the TIC of the warped sample with the TIC of the reference:

```
> sample2.dtw <- matrix(0, 100, 2000)
> for (i in 1:100) {
+    warpfun.dtw.i <- dtw(ssampM[i, ], srefM[i, ], keep = TRUE)
+    new.indices <- warp(warpfun.dtw.i, index.reference = FALSE)
+    sample2.dtw[i, ] <- ssampM[i, new.indices]
+ }
```

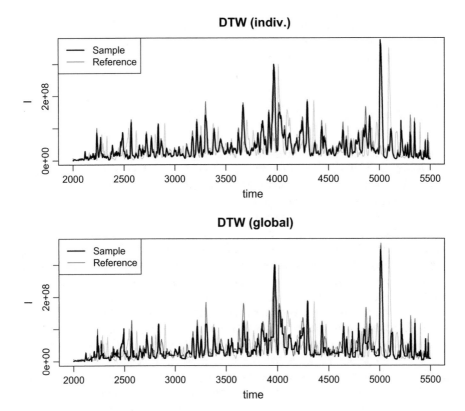

Fig. 3.13 TIC profiles of samples two and three of the `lcms` data. Top plot: DTW alignment using individual traces. Bottom plot: global DTW alignment

The result is shown in the top plot of Fig. 3.13—this should be compared with the top plot in Fig. 3.10. Clearly, the DTW result is much better. Note that since there is no overall compression, the length of the warped sample equals the original length.

Global alignment, using one warping function for all traces simultaneously, is available using the following code:

```
> warp.dtw.gl <- dtw(t(ssampM), t(srefM))
> samp.aligned <- lcms[, warp(warp.dtw.gl), 3]
```

The result, although still very good, is less convincing than the sum of individual DTW alignments: although the features are still aligned correctly, there are many examples of peak deformations. Global alignment with DTW is more constrained and therefore is forced to make compromises.

One should be careful, however, when applying extremely flexible warping methods such as DTW to data sets in which not all peaks can be matched. In this particular example we have aligned replicate measurements, so in principle every peak should be present in all samples. In practice, one very often will want to align different samples and alignment methods can easily be led astray by the presence of extra peaks, especially when these are of high intensity.

3.3.3   Practicalities

In almost all cases, a set of signals should be aligned in such a way that all features of interest are at the same positions in every trace. One strategy is to use the column means of the data matrix as a reference. This is only possible with very small shifts and will lead to peak broadening. Simply taking a random record from the set as a reference is better but still may be improved upon—it usually pays to perform some experiments to see which reference would lead to the smallest distortion of the other signals, while still leading to good alignment. If the number of samples is not too large, one can perform all possible combinations and see which one comes out best.

Careful data pretreatment is essential—baselines may severely influence the results and should be removed before alignment. In fact, one of the motivations of the CODA algorithm is to select traces that do not contain a baseline (Windig et al. 1996). Another point of attention is the fact that features can have intensities differing several orders in magnitude. Often, the biggest gain in the alignment optimization is achieved by getting the prominent features in the right location. Sometimes, this dominance leads to suboptimal alignments. Also differences in intensity between sample and reference signals can distort the results. Methods to cope with these phenomena will be treated in Sect. 3.5. Finally, it has been shown that in some cases results can be improved when the signals are divided into segments which are aligned individually (Wang and Isenhour 1987). Especially with more constrained warping methods like PTW this adds flexibility, but again, there is a danger of warping too much and mapping features onto the wrong locations. Especially in cases where there may be differences between the samples (control versus diseased, for instance) there is a risk that a biomarker peak, present only in one of the two classes, is incorrectly aligned. This, again, is all the more probable when that particular peak has a high intensity.

Packages **dtw** and **ptw** are by no means alone in tackling alignment. We already mentioned the **VPdtw** package: in addition, several Bioconductor packages, such as **PROcess** and **xcms**, implement both general and more specific alignment procedures, in most cases for mass-spectrometry data or hyphenated techniques like LC-MS.

3.4   Peak Picking

Several of the problems associated with misalignment can be avoided if the spectra can be transformed into lists of features, a process that is also known as peak picking. The first question of course is: what is a peak, exactly? This depends on the spectroscopic technique—usually it is a local maximum in a more or less smooth curve. In NMR, for instance, peaks usually have a specific shape (a Lorentz line shape). This knowledge can be used to fit the peaks to the data, and also to give quality assessments of the features that are identified. In chromatography, peaks can be described by a modified normal distribution, where the modification is allowing for peak tailing and other experimental imperfections. In cases where we do not want to make assumptions about peak shape, we are forced to more crude methods, e.g., finding a list of local maxima. One simple way to do this is again to make use of the embed function that splits up the spectrum in many overlapping segments. For each segment, we can calculate the location of the local maximum, and eliminate those segments where the local maximum is at the beginning or at the end. A function implementing this strategy is given in the next piece of code:

```
> pick.peaks <- function(x, span) {
+    span.width <- span * 2 + 1
+    loc.max <- span.width + 1 -
+       apply(embed(x, span.width), 1, which.max)
+    loc.max[loc.max == 1 | loc.max == span.width] <- NA
+
+    pks <- loc.max + 0:(length(loc.max) - 1)
+    unique(pks[!is.na(pks)])
+ }
```

The span parameter determines the width of the segments: wider segments will cause fewer peaks to be found. Let us investigate the effect using the prostate data from Fig. 3.2:

```
> pks10 <- pick.peaks(rmeans, 10)
> plot(Prostate2000Raw$mz[3:248], rmeans, type = "l",
+       xlab = "m/z", ylab = "Response", main = "span = 10")
> abline(v = Prostate2000Raw$mz[pks10 + 2], col = 2)
> pks40 <- pick.peaks(rmeans, 40)
> plot(Prostate2000Raw$mz[3:248], rmeans, type = "l",
+       xlab = "m/z", ylab = "Response", main = "span = 40")
> abline(v = Prostate2000Raw$mz[pks40 + 2], col = 2)
```

This leads to the plots in Fig. 3.14; with the wider span, many of the smaller features are not detected. At the same time, the many noisy features that are found with the smaller span, e.g., around m/z value 2040, probably do not constitute valid features. Clearly, the results of peak picking depend crucially on the degree and quality of the smoothing.

Once the positions of the features have been identified, one should quantify the signals. If an explicit peak model has been fitted, the normal approach would be to use the peak area, obtained by integrating between certain limits; if not, very often the peak height is taken. Under the assumption that peak widths are relatively constant, the two measures lead to similar results. If possible, one should then identify the

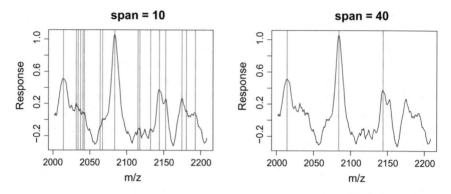

Fig. 3.14 Peak picking by identifying local maxima (prostate data): in the left figure the span is ten points, in the right forty

signals: in, e.g., mass spectrometry this would mean the identification of the corresponding fragment ion. Having these assignments makes it much easier to compare spectra of different samples: even if the features are not exactly at the same position, it still is clear which signals to compare. Thus, the need for peak alignment is obviated. In practice, however, it is rare to have complete assignments of spectral data from complex samples, and an alignment step remains necessary.

3.5 Scaling

The scaling method that is employed can totally change the result of an analysis. One should therefore carefully consider what scaling method (if any) is appropriate. Scaling can serve several purposes. Many analytical methods provide data that are not on an absolute scale; the raw data in such a case cannot be used directly when comparing different samples. If some kind of internal standard is present, it can be used to calibrate the intensities. In NMR, for instance, the TMS (tetramethylsilane, added to indicate the position of the origin on the x-axis) peak can be used for this if its concentration is known. Peak heights can then be compared directly. However, even in that situation it may be necessary to further scale intensities, since samples may contain different concentrations. A good example is the analysis of a set of urine samples by NMR. These samples will show appreciable global differences in concentrations, perhaps due to the amount of liquid the individuals have been consuming. This usually is not of interest—rather, one seeks one or perhaps a couple of metabolites with concentrations that deviate from the general pattern. As an example, consider the first ten spectra of the `prostate` data:

```
> range(apply(prostate[1:10, ], 1, max))
[1] 16.360 68.898
> range(rowSums(prostate[1:10, ]))
[1]  2531.7 15207.9
```

The intensity differences within these first ten spectra are already a factor five for both statistics. If these differences are not related to the phenomenon we are interested in but are caused, e.g., by the nature of the measurements, then it is important to remove them. As stated earlier, also in cases where alignment is necessary, this type of differences between samples can hamper the analysis.

Several options exist to make peak intensities comparable over a series of spectra. The most often-used are *range scaling*, *length scaling* and *variance scaling*. In range scaling, one makes sure that the data have the same minimal and maximal values. Often, only the maximal value is considered important since for many forms of spectroscopy zero is the natural lower bound. Length scaling sets the length of each spectrum to one; variance scaling sets the variance to one. The implementation in R is easy. Here, these three methods are shown for the first ten spectra of the prostate data. Range scaling can be performed by

```
> prost10.rangesc <- sweep(prostate[1:10, ], MARGIN = 1,
+                          apply(prostate[1:10, ], 1, max),
+                          FUN = "/")
> apply(prost10.rangesc, 1, max)
 1  2  3  4  5  6  7  8  9 10
 1  1  1  1  1  1  1  1  1  1
> range(rowSums(prost10.rangesc))
[1] 103.33 220.73
```

The sweep function is very similar to apply—it performs an action for every row or column of a data matrix. The MARGIN argument states which dimension is affected. In this case the MARGIN = 1 indicates the rows; column-wise sweeping would be achieved with MARGIN = 2. The third argument is the statistic that is to be swept out, here the vector of the per-row maximal values. The final argument states how the sweeping is to be done. The default is to use subtraction; here we use division. Clearly, the differences between the spectra have decreased.

Length scaling is done by dividing each row by the square root of the sum of its squared elements:

```
> prost10.lengthsc <- sweep(prostate[1:10, ], MARGIN = 1,
+                           apply(prostate[1:10, ], 1,
+                                 function(x) sqrt(sum(x^2))),
+                           FUN = "/")
> range(apply(prost10.lengthsc, 1, max))
[1] 0.11075 0.20581
> range(rowSums(prost10.lengthsc))
[1] 18.937 30.236
```

The difference between the smallest and largest values is now less than a factor of two. Scaling on the basis of variance or standard deviation has a similar effect:

```
> prost10.varsc <- sweep(prostate[1:10, ], MARGIN = 1,
+                          apply(prostate[1:10, ], 1, sd),
+                          FUN = "/")
> range(apply(prost10.varsc, 1, max))
[1] 11.697 21.659
> range(rowSums(prost10.varsc))
[1] 1976.5 3245.7
```

The underlying hypothesis in these scaling methods is that the maximal intensities, or the vector lengths, or the variances, should be equal in all objects. However, there is no real way of assessing whether these assumptions are correct, and it is therefore always advisable to assess different options.

Often in statistical modelling, especially in a regression context, we are more interested in deviations from a mean value than in the values per se. These deviations can be obtained by *mean-centering*, where one subtracts the mean value from every column in the data matrix, for example with the gasoline data:

```
> NIR.mc <- t(sweep(gasoline$NIR, 2, colMeans(gasoline$NIR)))
```

Subtraction is the default operation in `sweep`, so we do not need to provide that explicitly, but one can also use `sweep` to perform other functions by providing a `FUN =` argument. An even easier way to achieve the same effect is to use the `scale` function:

```
> NIR.mc <- scale(gasoline$NIR, scale = FALSE)
> matplot(wavelengths, t(NIR.mc),
+         type = "l", xlab = "Wavelength (nm)",
+         ylab = "1/R (mean-centered)", lty = 1)
```

The result is shown in Fig. 3.15; note the differences with the raw data shown in Fig. 2.1 and the first derivatives in Fig. 3.5.

When variables have been measured in different units or have widely different scales, we should take this into account. Obviously, one does not want the scale in which a variable is measured to have a large influence on the model: just switching to other units would lead to different results. One popular way of removing this dependence on units is called *autoscaling*, where every column x_i is replaced by

$$(x_i - \hat{\mu}_i)/\hat{\sigma}_i$$

In statistics, this is often termed *standardization* of data; the effect is that all variables are considered equally important. This type of scaling is appropriate for the wine data, since the variables have different units and very different ranges:

Fig. 3.15 Mean-centered
gasoline NIR data

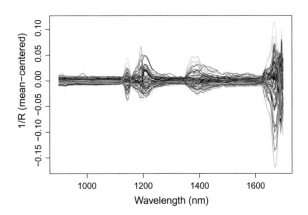

```
> apply(wines, 2, range)
      alcohol malic acid  ash ash alkalinity magnesium
[1,]    11.03       0.74 1.36            10.6        70
[2,]    14.83       5.80 3.23            30.0       162
      tot. phenols flavonoids non-flav. phenols proanth
[1,]          0.98       0.34             0.13    0.41
[2,]          3.88       5.08             0.66    3.58
      col. int. col. hue OD ratio proline
[1,]       1.28     0.48     1.27     278
[2,]      13.00     1.71     4.00    1680
```

The `apply` function in this case returns the range of every column. Clearly, the
last variable (proline) has values that are quite a lot bigger than the other variables.
The `scale` function already mentioned in the context of mean-centering also does
autoscaling—simply use the argument `scale = TRUE`, (the default, as it happens):

```
> wines.mc <- scale(wines, scale = FALSE)
> wines.sc <- scale(wines, scale = TRUE)
> boxplot(wines.mc ~ col(wines.mc),
+         main = "Mean-centered wine data")
> boxplot(wines.sc ~ col(wines.sc),
+         main = "Autoscaled wine data")
```

The result is shown in Fig. 3.16. In the left plot, showing the mean-centered data,
the dominance of proline is clearly visible, and any structure that may be present
in the other variables is hard to detect. The right plot is much more informative.
In almost all cases where variables indicate concentrations or amounts of chemical
compounds, or represent measurements in unrelated units, autoscaling is a good idea.
For the wine data, we will use always autoscaled data in examples, even in cases
where scaling does not matter.

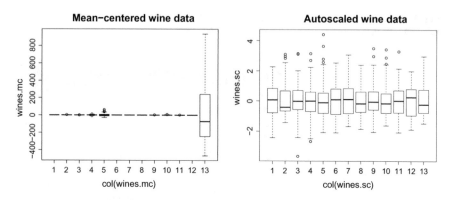

Fig. 3.16 Boxplots for the thirteen mean-centered variables (left) and auto-scaled variables (right) in the wine data set

For spectral data, prevalent in the natural sciences and the life sciences, on the other hand, autoscaling is usually not recommended. Very often, the data consist of areas of high information content, viz. containing peaks of different intensities, and areas containing only noise. When every spectral variable is set to the same standard deviation, the noise is blown up to the same size as the signal that contains the actual information. Clearly, this is not a desirable situation, and in such cases mean-centering is much preferred.

Specialized preprocessing methods are used a lot in spectroscopy. When the total intensity in the spectra is sample-dependent, spectra should be scaled in such a way that intensities can be compared. That is, one should scale the *rows* of the matrix (and after that, perhaps, also the columns). A typical example is given by the analysis of urine spectra (of whatever type): depending on how much a person has been drinking, urine samples can be more or less diluted, and therefore there is no direct relation between peak intensities in spectra from different subjects. This is a tricky phenomenon where domain knowledge should be used to decide on a sensible form of scaling. Sometimes it is known that one particular variable does not change much in different samples—in the urine context one could hypothesis that some component is always excreted at the same rate in all different persons. Such a compound could then be used as the yardstick and could be set to the same value (say, 100) in all samples by multiplication with an appropriate factor. The same factor then is applied to all other variables. An alternative is to scale to a particular total sum of responses. This, however, leads to so-called compositional data that in general require special data analysis approaches to avoid spurious correlations (Filzmoser et al. 2018).

One other form of row scaling is often used, especially in NIR applications: Standard Normal Variate scaling (SNV). This method essentially does autoscaling on the rows instead of the columns. That is, every spectrum will after scaling have a mean of zero and a standard deviation of 1. This gets rid of arbitrary offsets and multiplication factors. Obviously, the assumption that all spectra should have the same mean and variance is not always realistic! In some cases, the fact that the overall intensity in one spectrum is consistently higher may contain important information. In most cases, SNV gives results that are very close to MSC.

When noise is multiplicative in nature rather than additive, the level of variation depends on the signal strength. An example is mass spectrometry, where the error in signal intensity is often proportional to the intensity itself, and errors are routinely reported as relative standard deviations (the standard deviation divided by the absolute value of the mean). Since most noise reduction methods assume additive noise, a simple solution is to perform a log-transformation of the data. This decreases the influence of the larger features, and makes the noise more constant over the whole range. In addition, regular noise reduction methods can be applied if required. Log transformation can also be used to reduce the dominance of the largest features, which can be disturbing the analysis, e.g., in alignment applications. An often-used scaling with a similar albeit less pronounced effect is square-root scaling, which also has the advantage that it can be performed for data containing zeros. Also *Pareto scaling* decreases the effect of high-intensity variables, but not as dramatically as autoscaling or log scaling. It is basically autoscaling using the square root of the standard deviation in rescaling, rather than the standard deviation itself. This form of scaling has become popular in biomarker identification applications, e.g., in metabolomics: one would like to find variables that behave differently in two populations, and one is mostly interested in those variables that have high intensities (see also Sect. 11.5).

Deciding on an appropriate scaling is one of the most important decisions for many multivariate analyses, something that is made even more difficult by the fact that often several scaling steps are performed in sequence. Usually it is unclear what exactly the optimal scaling, or sets of scaling operations is—a few rules of thumb exist, but they are surely not covering all possible situations. Here is a stab at a list of the most common points, to be considered in sequence:

- a comparison between samples should be meaningful. That means that irrelevant concentration differences (as in urine samples) or irrelevant offsets (as in NIR spectra) should be removed as a first step.
- data based on counts often show an intensity-dependent error (often the relative standard deviation is constant)—in such cases, log-scaling is appropriate.
- for latent-variable methods like PCA, mean-centering is a must. Sometimes this is done automatically, but be sure to check.
- if all variables are thought to be equally important, at least in principle, then autoscaling may be appropriate. Note that for spectral data this often leads to increased noise in variables that do not contain information. In cases where one would like to concentrate on variables with a large signal, Pareto scaling may be preferable.

A good strategy determines before analysing the data how to do scaling; of course, one should not be afraid to change things if the scaling leads to unexpected disadvantages, but blindly trying several possible combinations and then selecting the one that seems most promising is bound to lead to problems. More on this topic can be found in Chap. 9 on validation.

3.6 Missing Data

Missing data are measurements that for some reason have not led to a valid result. In spectroscopic measurements, missing data are not usually encountered, but in many other areas of science they occur frequently. The main question to be answered is: are the data missing at random? If yes, then we can probably get around the problem, provided there are not too many missing data. If not, then it means that there is some rationale behind the missingness of data points. If we would know what it was, we could use that to decide how to handle the missing values. Usually, we don't, and that means trouble: if the missingness is related to the process we are studying our results will be biased and we can never be sure we are drawing correct conclusions.

Missing values in R are usually indicated by NA. Since many types of analysis do not accept data containing NAs, it is necessary to think of how to handle the missing values. If there are only a few, and they occur mostly in one or a few samples or one or a few variables, we might want to leave out these samples (or variables). Especially when the data set is rather large this seems a small price to pay for the luxury of having complete freedom in choosing any analysis method that is suited to our aim. Alternatively, one can try to find suitable replacements for the missing values, e.g. by estimating them from the other data points, a process that is known as *imputation*. Intermediate approaches are also possible, in which variables or samples with too many missing values are removed, and others, with a lower fraction of missing data, are retained. Sometimes imputation is not needed for statistical analysis: fitting linear models with lm for instance is possible also in the presence of missing values—these data points will simply be ignored in the fit process. Other functions such as var and cor have arguments that define several ways of dealing with missing values. In var, the argument is na.rm, allowing the user to either throw out missing values or accept missing values in the result, whereas cor has a more elaborate mechanism of defining strategies to deal with missing values. For instance, one can choose to consider only complete cases, or use only pairwise complete observations. Consult the manual pages for more information and examples. One example of dealing with missing values is shown in Sect. 11.1.

3.7 Conclusion

Data preprocessing is an art, in most cases requiring substantial background knowledge. Because several steps are taken sequentially, the number of possible schemes is often huge. Should one scale first and then remove noise or the other way around? Individual steps will influence each other: noise removal may make it more easy to correct for a sloping baseline, but the presence of a baseline may also influence your estimate of what is noise. General recipes are hard to give, but some problems are more serious than others. The presence of peak shifts, for instance, will make any multivariate analysis very hard to interpret.

Finally, one should realize that bad data preprocessing can never be compensated for in the subsequent analysis. One should *always* inspect the data before and after preprocessing and assess whether the relevant information has been kept while disturbing signals have been removed. Of course, that is easier said than done—and probably one will go through a series of modelling cycles before one is completely satisfied with the result.

Part II
Exploratory Analysis

Chapter 4
Principal Component Analysis

Principal Component Analysis or PCA (Jackson 1991; Jolliffe 1986) is a technique which, quite literally, takes a different viewpoint of multivariate data. It has many uses, perhaps the most important of which is the possibility to provide simple two-dimensional plots of high-dimensional data. This way, one can easily assess the presence of grouping or outliers, and more generally obtain an idea of how samples and variables relate to each other. PCA defines new variables, consisting of linear combinations of the original ones, in such a way that the first axis is in the direction containing most variation. Every subsequent new variable is orthogonal to previous variables, but again in the direction containing most of the remaining variation. The new variables are examples of what is called *latent variables* (LVs)—in the context of PCA the term *principal components* (PCs) is used.

The central idea is that more often than not many of the variables in high-dimensional data are superfluous. If we look at high-resolution spectra, for example, it is immediately obvious that neighboring wavelengths are highly correlated and contain similar information. Of course, one can try to pick only those wavelengths that appear to be informative, or at least differ from the other wavelengths in the selected set. This could, e.g., be based on clustering the variables, and selecting for each cluster one "representative". However, this approach is quite elaborate and will lead to different results when using different clustering methods and cutting criteria. Another approach is to use variable selection, given some criterion—one example is to select a limited set of variables leading to a matrix with maximal rank. Variable selection is notoriously difficult, especially in high-dimensional cases. In practice, many more or less equivalent solutions exist, which makes the interpretation quite difficult. We will come back to variable selection methods in Chap. 10.

PCA is an alternative. It provides a direct mapping of high-dimensional data into a lower-dimensional space containing most of the information in the original data. The tacit assumption here is that variation equals information. This is not always true, since variation may also be totally meaningless, e.g., in the case of noise.[1] The

---

[1] The reverse *is* true, however: if there is no variation in the data, there is no information either.

© Springer-Verlag GmbH Germany, part of Springer Nature 2020
R. Wehrens, *Chemometrics with R*, Use R!,
https://doi.org/10.1007/978-3-662-62027-4_4

coordinates of the samples in the new space are called *scores*, often indicated with the symbol T. The new dimensions are linear combinations of the original variables, and are called *loadings* (symbol P). The term Principal Component (PC) can refer to both scores and loadings; which is meant is usually clear from the context. Thus, one can speak of sample coordinates in the space spanned by PC 1 and 2, but also of variables contributing greatly to PC 1.

The matrix multiplication of scores and loadings leads to an approximation $\tilde{X}$ of the original data X:

$$X = \tilde{X} + E = T_a P_a^T + E \tag{4.1}$$

Superscript ^{T}, as usual, indicates the transpose of a matrix. The subscript a indicates how many components are taken into account. Taking more numbers into account will improve the agreement between X and $\tilde{X}$ and decrease the size of the elements of the residual matrix E. The largest possible number of PCs is the minimum of the number of rows and columns of the matrix:

$$a_{max} = \min(n, p) \tag{4.2}$$

If $a = a_{max}$, the approximation is perfect and $\tilde{X} = X$.

The PCs are orthogonal combinations of variables defined in such a way that (Jolliffe 1986):

- the variances of the scores are maximal;
- the sum of the Euclidean distances between the scores is maximal;
- the reconstruction of X is as close as possible to the original: $||X - \tilde{X}||$ is minimal.

These three criteria are equivalent (Jackson 1991); the next section will show how to find the PCs.

PCA has many advantages: it is simple, has a unique analytical solution optimizing a clear and unambiguous criterion, and often leads to a more easily interpretable data representation. The price we have to pay is that we do not have a small set of wavelengths carrying the information but a small set of principal components, in which *all* wavelengths are represented. Note that the underlying assumption is that variation equals information. Intuitively, this makes sense: one can not learn much from a constant number.

Once PCA has defined the latent variables, one can plot all samples in the data set while ignoring all higher-order PCs. Usually, only a few PCs are needed to capture a large fraction of the variance in the data set (although this is highly dependent on the type of data). That means that a plot of (the scores of) PC 1 versus PC 2 can already be highly informative. Equally useful is a plot of the contributions of the (original) variables to the important PCs. These visualizations of high-dimensional data perhaps form the most important application of PCA. Later, we will see that the scores can also be used in regression and classification. Also the residual matrix E can contain useful information. We'll see two examples in Section 11.2, on robust PCA, and Section 11.3 on multivariate process control.

4.1 The Machinery

Currently, PCA is implemented even in low-level numerical software such as spreadsheets. Nevertheless, it is good to know the basics behind the computations. In almost all cases, the algorithm used to calculate the PCs is *Singular Value Decomposition* (SVD).[2] It decomposes an $n \times p$ mean-centered data matrix X into three parts:

$$X = UDV^T \tag{4.3}$$

where U is a $n \times a$ orthonormal matrix containing the left singular vectors, D is a diagonal matrix $(a \times a)$ containing the singular values, and V is a $p \times a$ orthonormal matrix containing the right singular vectors. The latter are what in PCA terminology is called the loadings—the product of the first two matrices forms the scores:

$$X = (UD)V^T = TP^T \tag{4.4}$$

The interpretation of matrices T, P, U, D and V is straightforward. The loadings, columns in matrix P (or equivalently, the right singular vectors, columns in matrix V) give the weights of the original variables in the PCs. Variables that have very low values in a specific column of V contribute only very little to that particular latent variable. The scores, columns in T, constitute the coordinates in the space of the latent variables. Put differently: these are the coordinates of the samples as we see them from our new PCA viewpoint. The columns in U give the same coordinates in a normalized form—they have unit variances, whereas the columns in T have variances corresponding to the variances of each particular PC. These variances λ_i are proportional to the squares of the diagonal elements in matrix D:

$$\lambda_i = d_i^2/(n-1)$$

The fraction of variance explained by PC i can therefore be expressed as

$$FV(i) = \lambda_i/\sum_{j=1}^{a}\lambda_j \tag{4.5}$$

One main problem in the application of PCA is the decision on how many PCs to retain; we will come back to this in Section 4.3.

One final remark needs to be made about the unique solution given by the SVD algorithm: it is only unique up to the sign. As is clear from, e.g., Equation 4.4, one can obtain exactly the same solution by reversing the sign of *both* scores and loadings

---

[2] One alternative for SVD is the application of the Eigen decomposition on the covariance or correlation matrix of the data. SVD is numerically more stable and is therefore preferred in most cases.

simultaneously. There are no conventions, so one should always keep in mind that this possibility exists. For the interpretation of the data, it make no difference whatsoever.

Although SVD is a fast algorithm in some cases it can be efficient not to apply it to the data matrix directly, especially in cases where there is a large difference in the numbers of rows and columns. In such a case, it is faster to apply SVD to either $X^T X$ or XX^T, whichever is the smaller one. If the number of columns is much smaller than the number of rows, one would obtain

$$X^T X = (UDV^T)^T UDV^T = VD^2 V^T$$

Applying svd[3] directly yields loadings and sums of squares. Matrix T, the score matrix, is easily found by right-multiplying both sides of Equation 4.4 with P:

$$XP = TP^T P = T \tag{4.6}$$

because of the orthonormality of P. Similarly, we can find the left singular vectors and singular values when applying SVD to XX^T—see the example in the next section.

4.2 Doing It Yourself

Calculating scores and loadings is easy: consider the wine data first. We perform PCA on the autoscaled data to remove the effects of the different scales of the variables using the svd function provided by R:

```
> wines.svd <- svd(wines.sc)
> wines.scores <- wines.svd$u %*% diag(wines.svd$d)
> wines.loadings <- wines.svd$v
```

The first two PCs represent the plane that contains most of the variance; how much exactly is given by the squares of the values on the diagonal of D. The importance of individual PCs is usually given by the percentage of the overall variance that is explained:

```
> wines.vars <- wines.svd$d^2 / (nrow(wines) - 1)
> wines.totalvar <- sum(wines.vars)
> wines.relvars <- wines.vars / wines.totalvar
> variances <- 100 * round(wines.relvars, digits = 3)
> variances[1:5]
[1] 36.0 19.2 11.2  7.1  6.6
```

The first PC covers more than one third of the total variance; for the fifth PC this amount is down to one fifteenth.

---

[3]Or eigen, which returns eigenvectors and eigenvalues.

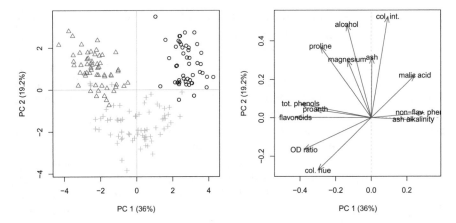

Fig. 4.1 Left plot: scores on PCs 1 and 2 for the autoscaled wine data. Different symbols correspond to the three cultivars. Right plot: loadings on PCs 1 and 2

The scores show the positions of the individual wine samples in the coordinate system of the PCs. A score plot can be produced as follows:

```
> plot(wines.scores[, 1:2], type = "n",
+       xlab = paste("PC 1 (", variances[1], "%)", sep = ""),
+       ylab = paste("PC 2 (", variances[2], "%)", sep = ""))
> abline(h = 0, v = 0, col = "gray")
> points(wines.scores[, 1:2], pch = wine.classes, col = wine.classes)
```

The result is depicted in the left plot of Fig. 4.1. It is good practice to indicate the amount of variance associated with each PC on the axis labels. The three cultivars can be clearly distinguished: class 1, Barbera, indicated with black open circles, has the largest scores on PC 1 and class 2 (Barolo—red triangles in the figure) the smallest. PC 2, corresponding to 19% of the variance, improves the separation by separating the Grignolinos, the middle class on PC 1, from the other two. Note that this is a happy coincidence: PCA does not explicitly look to discriminate between classes. In this case, the three cultivars clearly have different characteristics. What characteristics these are can be seen in the loading plot, shown on the right in Fig. 4.1. It shows the contribution of the original variables to the PCs. Loadings are traditionally shown as arrows from the origin:

```
> plot(wines.loadings[, 1] * 1.2, wines.loadings[, 2], type = "n",
+       xlab = paste("PC 1 (", variances[1], "%)", sep = ""),
+       ylab = paste("PC 2 (", variances[2], "%)", sep = ""))
> abline(h = 0, v = 0, col = "gray", lty = 2)
> arrows(0, 0, wines.loadings[, 1], wines.loadings[, 2],
+       col = "blue", length = .15, angle = 20)
> text(wines.loadings[, 1:2], labels = colnames(wines))
```

The factor of 1.2 in the plot command is used to create space for the text labels. Clearly, the wines of class 3 (green symbols in the plot) are distinguished by lower values of alcohol and a lower color intensity. Wines of class 2 (in red) have high flavonoid and phenol content and are low in non-flavonoid phenols; the reverse is true for wines of class 1. All of these conclusions could probably have been drawn also by looking at class-specific boxplots for all variables—however, the combination of one score plot and one loading plot shows this in a much simpler way, and even presents direct information on correlations between variables and objects. We will come back on this point later, when treating biplots.

As an example of the kind of speed improvement one can expect when applying SVD on the crossproduct matrices rather than the original data, consider a really wide matrix with 100 rows and 100,000 columns (what the values in the matrix are does not matter for the timing). Timings can be obtained by wrapping the code within a system.time call:

```
> nr <- 100
> nc <- 100000
> X <- matrix(rnorm(nr*nc), nrow = nr)
> system.time({
+     X.svd <- svd(X)
+     X.scores <- X.svd$u %*% diag(X.svd$d)
+     X.variances <- X.svd$d^2 / (nrow(X) - 1)
+     X.loadings <- X.svd$v
+ })
   user  system elapsed
  3.715   0.099   3.815
```

Here, the number of variables is much larger than the number of objects (which, by the way, is not extremely small either), so we perform SVD on the matrix XX^T:

```
> system.time({
+     X2.tcp <- tcrossprod(X)
+     X2.svd <- svd(X2.tcp)
+     X2.scores <- X2.svd$u %*% diag(sqrt(X2.svd$d))
+     X2.variances <- X2.svd$d / (nrow(X) - 1)
+     X2.loadings <- solve(X2.scores, X)
+ })
   user  system elapsed
  1.709   0.012   1.721
```

The second option is more than twice as fast—for bigger data sets, or in cases where many PCAs need to be done, this may become very relevant. Instead of the crossproduct matrix, also the covariance matrix or correlation matrix can be used in exactly the same way—using the correlation matrix corresponds to performing PCA on autoscaled data.

4.3 Choosing the Number of PCs

The question how many PCs to consider, or put differently: where the information stops and the noise begins, is difficult to answer. Many methods consider the amount of variance explained, and use statistical tests or graphical methods to define which PCs to include. In this section we briefly review some of the more popular methods.

4.3.1 Scree Plots

The amount of variance per PC is usually depicted in a scree plot: either the variances themselves or the logarithms of the variances are shown as bars. Often, one also considers the fraction of the total variance explained by every single PC. The last few PCs usually contain no information and, especially on a log scale, tend to make the scree plot less interpretable, so they are usually not taken into account in the plot.

```
> barplot(wines.vars[1:10], main = "Variances",
+         names.arg = paste("PC", 1:10))
> barplot(log(wines.vars[1:10]), main = "log(Variances)",
+         names.arg = paste("PC", 1:10))
> barplot(wines.relvars[1:10], main = "Relative variances",
+         names.arg = paste("PC", 1:10))
> barplot(cumsum(100 * wines.relvars[1:10]),
+         main = "Cumulative variances (%)",
+         names.arg = paste("PC", 1:10), ylim = c(0, 100))
```

This leads to the plots in Fig. 4.2. Clearly, PCs 1 and 2 explain much more variance than the others: together they cover 55% of the variance. The scree plots show no clear cut-off, which in real life is the rule rather than the exception. Depending on the goal of the investigation, for these data one could consider three or five PCs. Choosing four PCs would not make much sense in this case, since the fifth PC would explain almost the same amount of variance: if the fourth is included, the fifth should be, too.

4.3.2 Statistical Tests

One can show that the explained variance for the i-th component, $\lambda_i = d_i^2/(n-1)$ is asymptotically normally distributed (Härdle and Simar 2007; Mardia et al. 1979), which leads to confidence intervals of the form

$$\ln(\lambda_i) \pm z_\alpha \sqrt{\frac{2}{n-1}}$$

For the wine example, 95% confidence intervals can therefore be obtained as

```
> llambdas <- log(wines.vars)
> CIwidth <- qnorm(.975) * sqrt(2 / (nrow(wines) - 1))
> CIs <- cbind(exp(llambdas - CIwidth),
+              wines.vars,
+              exp(llambdas + CIwidth))
> colnames(CIs) <- c("CI 0.025", "  Estimate", "  CI 0.975")
> CIs[1:4, ]
      CI 0.025    Estimate    CI 0.975
[1,]   3.79580    4.67780      5.7647
[2,]   2.02973    2.50137      3.0826
[3,]   1.17930    1.45333      1.7910
[4,]   0.75014    0.92444      1.1393
```

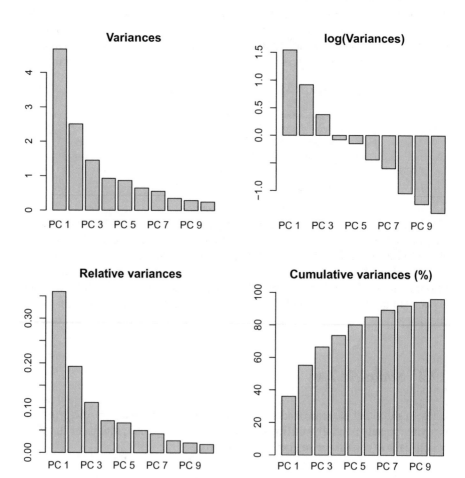

Fig. 4.2 Scree plots for the assessment of the amount of variance explained by each PC. From left to right, top to bottom: variances, logarithms of variances, fractions of the total variance and cumulative percentage of total variance

Mardia et al. present an approach testing the equality of variances for individual PCs (Mardia et al. 1979). For the autoscaled wine data, one could test whether the last $p - k$ PCs are equally important, i.e., have equal values of λ. The quantity

$$(n - 1)(p - k) \log(a_0/g_0)$$

is distributed approximately as a χ^2-statistic with $(p - k + 2)(p - k - 1)/2$ degrees of freedom (Mardia et al. 1979, p. 236). In this formula, a_0 and g_0 indicate arithmetic and geometric means of the $p - k$ smallest variances, respectively. We can use this test to assess whether the last three PCs are useful or not:

```
> small.ones <- wines.vars[11:13]
> n <- nrow(wines)
> nsmall <- length(small.ones)
> geo.mean <- prod(small.ones)^{1/nsmall}
> mychisq <- (n - 1) * nsmall * log(mean(small.ones) / geo.mean)
> ndf <- (nsmall + 2) * (nsmall - 1) / 2
> 1 - pchisq(mychisq, ndf)
[1] 9.9111e-05
```

This test finds that after PC 10 there is still a difference in the variances. In fact, the test finds a difference after any other cutoff, too: apparently all PCs are significant.

The use of statistical tests for determining the optimal number of PCs has never really caught on. Most scientists are prepared to accept a certain loss of information (variance) provided that the results, the score plots and loading plots, help to answer scientific questions. Most often, one uses informal graphical methods: if an elbow shows up in the scree plot that can be used to decide on the number of PCs. In other applications, notably with spectral data, one can sometimes check the loadings for structure. If there is no more structure—in the form of peak-like shapes—present in the loadings of higher PCs, they can safely be discarded.

4.4 Projections

Once we have our low-dimensional view of the original high-dimensional data, we may be interested how other data are positioned. This may be data from new samples, measured on a different instrument, or on a different day. The key point is that the low-dimensional representation allows us to look at the data and in one glance assess whether there are patterns. Obtaining scores for new data is pretty easy. Given a new data matrix X, the projections in the space defined by loadings P can be obtained by simple right-multiplication:

$$XP = TP^T P = T$$

The wine data, for example, are more or less ordered according to vintage. If we would perform PCA on the first half of the data, the third class, the Barbera wines, would not play a part in defining the PCs at all, and the first class, Barolo, would dominate. To see the effect of this, we can project the second half of the data matrix into the PCA space defined by the first half. We start by constructing scores and loadings for the first half:

```
> X1 <- scale(wines[1:88, ])
> X1.svd <- svd(X1)
> X1.pca <- list(scores = X1.svd$u %*% diag(X1.svd$d),
+                loadings = X1.svd$v)
```

Then, we scale the second half of the data using the means and standard deviations of the first half. This is a very important detail that sometimes is missed—obviously, both halves should have the same point of origin:

```
> X2 <- scale(wines[89:177, ],
+             center = attr(X1, "scaled:center"),
+             scale = attr(X1, "scaled:scale"))
> X2.scores <- X2 %*% X1.pca$loadings
```

Showing the scores now is easy:

```
> labels <- rep(c(1, 2), c(89, 88))
> plot(rbind(X1.pca$scores, X2.scores),
+      pch = labels, col = labels,
+      xlab = "PC 1", ylab = "PC 2")
```

This leads to the left plot in Fig. 4.3. Clearly, the familiar shape of the PC 1 vs. PC 2 plot has been destroyed, and what is more: the triangles, corresponding to the second half of the data, are generally in a different location than the first half (circles). Since Barbera wines are only present in the second half, and Barolo wines only in the first half, this comes as no surprise. The right plot in the same figure shows a completely different picture. It has been obtained by defining X1 and X2 as follows:

```
> wines.odd <- seq(1, nrow(wines), by = 2)
> wines.even <- seq(2, nrow(wines), by = 2)
> X1b <- scale(wines[wines.odd, ])
> X2b <- scale(wines[wines.even, ],
+             center = attr(X1b, "scaled:center"),
+             scale = attr(X1b, "scaled:scale"))
```

Now both halves have similar compositions, and the data clouds neatly overlap. In fact, this is a very important way to check whether a division in training and test set, a topic that we will talk about extensively in later chapters, is a good one.

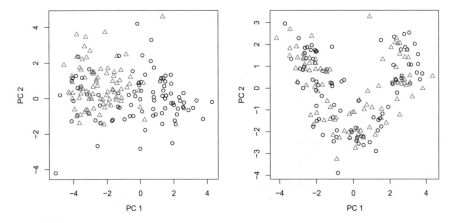

Fig. 4.3 Projections in PCA space. Left plot: second half of the wine data (red triangles) projected into the PCA space defined by the first half (black circles). Right plot: PCA model based on the odd rows (circles). The even rows (triangles) are projected in this space. The result is very similar to a PCA on the complete data matrix

4.5 R Functions for PCA

The standard R function for PCA is prcomp. By default, the data are mean-centered (but not scaled!). We will show its use on the gasoline data, limiting the number of PCs to six:

```
> nir.prcomp <- prcomp(gasoline$NIR, rank. = 6)
> summary(nir.prcomp)
Importance of first k=6 (out of 60) components:
                          PC1     PC2     PC3     PC4     PC5     PC6
Standard deviation      0.210  0.0831  0.0651  0.0529  0.0275  0.02426
Proportion of Variance  0.726  0.1134  0.0695  0.0460  0.0124  0.00967
Cumulative Proportion   0.726  0.8390  0.9086  0.9546  0.9670  0.97664
```

One can see that these six components explain almost 98% of the variance. The default plot command is to show a scree plot (in fact, the screeplot function is used):

```
> plot(nir.prcomp, main = "Gasoline scree plot")
```

The result is depicted in Fig. 4.4. Probably, most people would select four components to be included: although the first is much larger than the others, components two to four still contribute a substantial amount, whereas higher components are much less important.

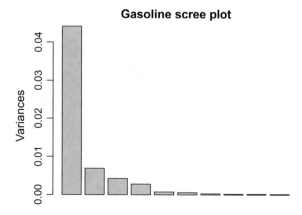

Fig. 4.4 Scree plot for the NIR data. By default, the scree plot shows not more than ten PCs

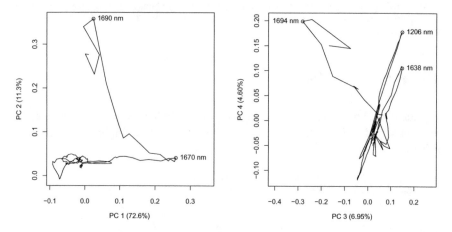

Fig. 4.5 Loading plots for the mean-centered gasoline data; the left plot shows PC 1 versus PC 2, and the right plot PC 3 versus PC 4. Some extreme variable weights are indicated with the corresponding wavelengths

Plotting the loadings of the first four PCs shows some interesting structure. The loadings are available as the `rotation` element of the `prcomp` object:

```
> nir.loadings <- nir.prcomp$rotation[, 1:4]
```

The original variables are, of course, highly correlated, and therefore connecting subsequent variables in the loading space forms a trajectory. In the following code, producing the left plot (PC 1 against PC 2) in Fig. 4.5, the variables with the most extreme loadings have been indicated—these can easily be found using the `identify` function (see the manual page for details). The plot for PCs 3 and 4 is made completely analogously.

```
> offset <- c(0, 0.09) # to create space for labels
> plot(nir.loadings[, 1:2], type = "l",
+      xlim = range(nir.loadings[, 1]) + offset,
+      xlab = "PC 1 (72.6%)", ylab = "PC 2 (11.3%)")
> points(nir.loadings[c(386, 396), 1:2])
> text(nir.loadings[c(386, 396), 1:2], pos = 4,
+      labels = paste(c(1670, 1690), "nm"))
```

We can see that the variation on PC 1 is mainly attributable to the intensities around 1670 nm, whereas the wavelengths around 1690 nm are contributing most to PC 2. PC 3 shows the largest loadings at 1638, 1694 and 1206 nm; the latter two are also extreme loadings on PC 4. These wavelengths correspond with areas of significant variation (see Fig. 2.1).

An even more interesting visualization is the *biplot* (Gabriel 1971; Gower and Hand 1996). This shows scores and loadings in one and the same plot, which can make interpretation easier. The origins for the score and loading plots are overlayed, and the two sets of points are plotted in separate axis systems. An example is shown in Fig. 4.6:

```
> biplot(nir.prcomp, col = c("black", "blue"), scale = 0.5)
```

Again, scores are plotted as individual points, and loadings as arrows. The axes corresponding to the scores are shown on the bottom and the left of the picture; the axes to the right and on top are for the loadings. Since there are many correlated variables, the loading plot looks quite crowded. The `scale` argument (taking values between 0 and 1) can be used to reach a visually pleasing balance between the sizes

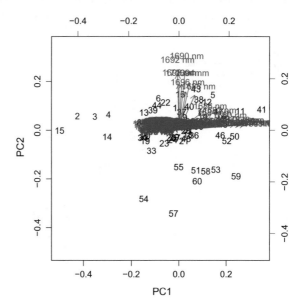

Fig. 4.6 Biplot for the mean-centered gasoline data

of the scores and the loadings. The setting employed here uses square-root scaling for both variables and observations.

We can see in the biplot that wavelength 1690 nm is one of the arrows pointing upward, in the direction of PC 2. Samples 15 and 41 are most different in PC 1, and samples 57 and 43 differ most in PC 2. Let us plot the mean-centered data for these four samples (since the mean-centered data are what is fed to PCA):

```
> extremes <- c(15, 41, 43, 57)
> Xextr <- scale(gasoline$NIR, scale = FALSE)[extremes, ]
> matplot(wavelengths, t(Xextr),
+         type = "l", xlab = "Wavelength (nm)",
+         ylab = "Intensity (mean-scaled)", lty = c(1, 1, 2, 2),
+         col = 1:4)
> legend("bottomleft", legend = paste("sample", extremes),
+         lty = c(1, 1, 2, 2), col = 1:4, bty = "n")
```

This results in Fig. 4.7. The largest difference in intensity, at 1670 nm, corresponds with the most important variables in PC 1—and although it is not directly recognizable from the loadings in this figure, this is exactly the wavelength with the largest loading on PC 1. The largest difference in the samples that are extreme in PC 2 is just above that, at 1690 nm, in agreement with our earlier observation. Another major feature just below 1200 nm is represented in the bottom left corner of the loading plot of PC 1 vs PC 2.

Another R function, princomp, also does PCA, using the eigen decomposition on the covariance matrix instead of directly performing svd on the data themselves. Since the svd-based calculations are more stable, these are to be preferred for regular applications; princomp does allow you to provide a specific covariance matrix that will be used for the decomposition, which can be useful in some situations (see Section 11.2), but is retained mainly for compatibility reasons.

Fig. 4.7 Mean-centered spectra of samples that are extreme in PCs 1 or 2

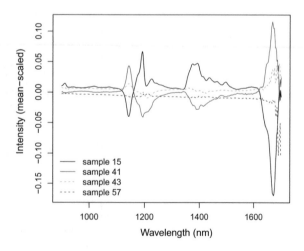

The package **ChemometricsWithR** comes with a set of PCA functions, too, based on the code presented in this chapter. The basic function is PCA, and the usual generic functions print, plot, and summary are available, as well as some auxiliary functions such as screeplot, project, and the extraction functions variances, loadings and scores. To produce, e.g., plots very similar to the ones shown in Fig. 4.1, one can issue:

```
> wines.PCA <- PCA(scale(wines))
> scoreplot(wines.PCA, pch = wine.classes, col = wine.classes)
> loadingplot(wines.PCA, show.names = TRUE)
```

One useful feature of these functions is that the percentages of explained variance are automatically shown at the axes.

4.6   Related Methods

PCA is not alone in its aim to find low-dimensional representations of high-dimensional data sets. Several other methods try to do the same thing, but rather than finding the projection that maximizes the explained variance, they choose other criteria. In Principal Coordinate Analysis (PCoA) and the related Multidimensional Scaling (MDS) methods, the aim is to find a low-dimensional projection that reproduces the experimentally found *distances* between the data points. When these distances are Euclidean, the results are the same or very similar to PCA results; however, other distances can be used as well. Independent Component Analysis maximizes deviations from normality rather than variance, and Factor Analysis concentrates on reproducing covariances. We will briefly review these methods in the next paragraphs.

4.6.1   Multidimensional Scaling

In some cases, applying PCA to the raw data matrix is not appropriate, for example in situations where regular Euclidean distances do not apply—similarities between chemical structures, e.g., can be expressed easily in several different ways, but it is not at all clear how to represent molecules into fixed-length structure descriptors (Baumann 1999), something that is required by distance measures such as the Euclidean distance. Even when comparing spectra or chromatograms, the Euclidean distance can be inappropriate, for instance in the presence of peak shifts (Bloemberg et al. 2010; de Gelder et al. 2001). In other cases, raw data are simply not available and the only information one has consists of similarities. Based on the sample similarities, the goal of methods like Multidimensional Scaling (MDS, (Borg and Groenen 2005; Cox and Cox 2001)) is to reconstruct a low-dimensional map of samples that leads to the same similarity matrix as the original data (or a very close approximation).

Since visualization usually is one of the main aims, the number of dimensions usually is set to two, but in principle one could find an optimal configuration with other dimensionalities as well.

The problem is something like making a topographical map, given only the distances between the cities in the country. In this case, an exact solution is possible in two dimensions since the original distance matrix was calculated from two-dimensional coordinates. Note that although distances can be reproduced exactly, the map still has rotational and translational freedom—in practice this does not pose any problems, however. An amusing example is given by maps not based on kilometers but rather on travel time—the main cities will be moved to the center of the plot since they usually are connected by high-speed trains, whereas smaller villages will appear to be further away. In such a case, and in virtually all practical applications, a two-dimensional plot will not be able to reproduce all similarities exactly.

In MDS, there are several ways to indicate the agreement between the two distance matrices, and these lead to different methods. The simplest approach is to perform PCA on the double-centered distance matrix,[4] an approach that is known as Principal Coordinate Analysis, or Classical MDS (Gower 1966). The criterion to be minimized is called the stress, and is given by

$$S = \sum_{j<i}(||x_i - x_j|| - e_{ij})^2 = \sum_{j<i}(d_{ij} - e_{ij})^2$$

where e_{ij} corresponds with the true, given, distances, and d_{ij} are the distances between objects x_i and x_j in the low-dimensional space.

In R, this is available as the function cmdscale:

```
> wines.dist <- dist(scale(wines))
> wines.cmdscale <- cmdscale(wines.dist)
> plot(wines.cmdscale,
+       pch = wine.classes, col = wine.classes,
+       main = "Principal Coordinate Analysis",
+       xlab = "Coord 1", ylab = "Coord 2")
```

This leads to the left plot in Fig. 4.8, which up to the reversal of the sign in the second component is exactly equal to the score plot of Fig. 4.1.

Other approaches optimize slightly different criteria: two well-known examples are Sammon mapping and Kruskal-Wallis mapping (Ripley 1996)—both are available in the **MASS** package as functions sammon and isoMDS, respectively. Sammon mapping decreases the influence of large distances, which can dominate the map completely. It minimizes the following stress criterion:

$$S = \frac{1}{\sum_i \sum_{j<i} d_{ij}} \sum_i \sum_{j<i} \frac{d_{ij} - e_{ij}}{d_{ij}}$$

---

[4]Double centering is performed by mean-centering in both row and column dimensions, and subsequently adding the grand mean of the original matrix to center the data around the origin.

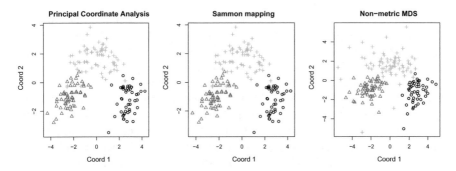

Fig. 4.8 Multidimensional scaling approaches on the wine data: classical MDS (left), Sammon mapping (middle) and non-metric MDS (right)

Since no analytical solution is available, gradient descent optimization is employed to find the optimum. The starting point usually is the classical solution, but one can also provide another configuration—indeed, one approach to try to avoid local optima is to repeat the mapping starting from many different starting sets, or to use different sets of search parameters. Sammon mapping is easy (the argument `trace = FALSE` suppresses progress output):

```
> wines.sammon <- sammon(wines.dist, trace = FALSE)
> plot(wines.sammon$points, main = "Sammon mapping",
+       col = wine.classes, pch = wine.classes,
+       xlab = "Coord 1", ylab = "Coord 2")
```

Note that the result is a list rather than a coordinate matrix: the coordinates in low-dimensional space can be accessed in list element `points`.

The non-metric scaling implemented in `isoMDS` uses a two-step optimization that alternatively finds a good configuration in low-dimensional space, and an appropriate non-monotone transformation. In effect, one finds a set of points that leads to the same *order* of the distances in the low-dimensional approximation and in the real data, rather than resulting in approximately the same distances.

```
> wines.isoMDS <- isoMDS(wines.dist, trace = FALSE)
> plot(wines.isoMDS$points, main = "Non-metric MDS",
+       col = wine.classes, pch = wine.classes,
+       xlab = "Coord 1", ylab = "Coord 2")
```

The results of Sammon mapping and IsoMDS are shown in the middle and right plots of Fig. 4.8, respectively. Here we again see the familiar horse-shoe shape, although it is somewhat different in nature in the non-metric version in the plot on the right. There, the Grignolino class is much more spread out than the other two.

MDS is popular in the social sciences, but much less so in the life sciences: maybe there its disadvantages are more important. The first drawback is that MDS requires a full distance matrix. For data sets with many thousands of samples this can be prohibitive in terms of computer memory. The other side of the coin is that the (nowadays more common) data sets with many more samples than variables do not present any problem; they can be analyzed by MDS easily. The second and probably more important disadvantage is that MDS does not provide an explicit mapping operator. That means that new objects cannot be projected into the lower-dimensional point configuration as we did before with PCA; either one redoes the MDS mapping, or one positions the new samples as good as possible within the space of the mapped ones and takes several iterative steps to obtain a new, complete, set of low-dimensional coordinates. Finally, the fact that the techniques rely on optimization, rather than an analytical solution, is a disadvantage: not only does it take more time, especially with larger data sets, but also the optimization settings may need to be tweaked for optimal performance.

4.6.2 Independent Component Analysis and Projection Pursuit

Variation in many cases equals information, one of the reasons behind the widespread application of PCA. Or, to put it the other way around, a variable that has a constant value does not provide much information. However, there are many examples where the *relevant* information is hidden in small differences, and is easily overwhelmed by other sources of variation that are of no interest. The technique of *Projection Pursuit* (Friedman 1987; Friedman and Tukey 1974; Huber 1985) is a generalization of PCA where a number of different criteria can be optimized. One can for instance choose a viewpoint that maximizes some grouping in the data. In general, however, there is no analytical solution for any of these criteria, except for the variance criterion used in PCA. A special case of Projection Pursuit is *Independent Component Analysis* (ICA, Hyvärinen et al. 2001), where the view is taken to maximize deviation from multivariate normality, given by the negentropy J. This is the difference of the entropy of a normally distributed random variable $H(x_G)$ and the entropy of the variable under consideration $H(x)$

$$J(x) = H(x_G) - H(x)$$

where the entropy itself is given by

$$H(x) = -\int f(x) \log f(x) dx$$

Since the entropy of a normally distributed variable is maximal, the negentropy is always positive (Cover and Thomas 1991). Unfortunately, this quantity is hard to calculate, and in practice approximations, such as kurtosis and the fourth moment are used. Package **fastICA** provides the fastICA procedure of Hyvarinen and Oja

ICA components

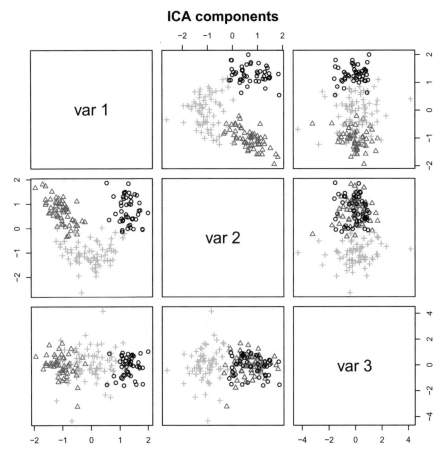

Fig. 4.9 Fast ICA applied to the wine data, based on a three-component model

(Hyvärinen and Oja 2000), employing other approximations that are more robust and faster. The algorithm can be executed either fully in R, or using C code for greater speed.

The fastICA algorithm first mean-centers the data, and then performs a "whitening", i.e., a Principal Component Analysis. The PCs are then rotated in order to optimize the non-gaussianity criterion. For the wine data, this gives the result shown in Fig. 4.9:

```
> wines.ica <- fastICA(wines.sc, 3)
> pairs(wines.ica$S, main = "ICA components",
+        col = wine.classes, pch = wine.classes)
```

Note that repeated application may lead to different local optima, and thus to different results. One additional characteristic that should be noted is that ICA components

ICA components (3 out of 5)

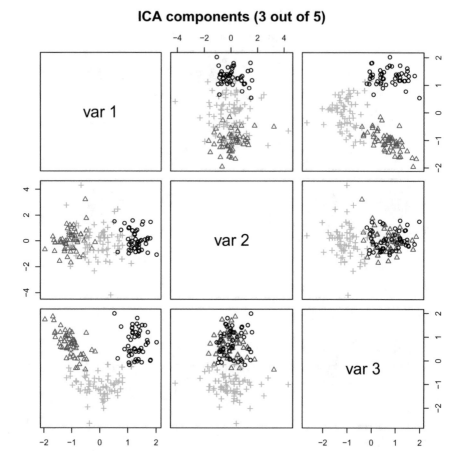

Fig. 4.10 Fast ICA applied to the wine data using five components; only the first three components are shown

can change, depending on the total number of components in the model: where in PCA the first component remains the same, no matter how many other components are included, in ICA this is not the case. The pairs plot for the first three components will therefore be different when taking, e.g., a five-component ICA model (shown in Fig. 4.10) compared to the three-component model in Fig. 4.9:

```
> wines.ica5 <- fastICA(wines.sc, 5)
> pairs(wines.ica5$S[, 1:3],
+       main = "ICA components (3 out of 5)",
+       col = wine.classes, pch = wine.classes)
```

In Fig. 4.10 it is interesting to note that the second ICA component is not related to the difference in variety. Instead, it is the plot of IC 1 versus IC 3 that shows most discrimination, and is most similar to the PCA score plot shown in Fig. 4.1.

Clearly, one should be cautious in the inspection of ICA score plots. Of course, class separation is not the only criterion we can apply—just because we happen to know in this case what the grouping is does not mean that all useful projections should be showing it!

4.6.3 Factor Analysis

Another procedure closely related to PCA is Factor Analysis (FA), developed some eighty years ago by the psychologist Charles Spearman, who hypothesized that a large number of abilities (mathematical, artistic, verbal) could be summarized in one underlying factor "intelligence" (Spearman 1904). Although this view is no longer mainstream, the idea caught on, and FA can be summarized as trying to describe a set of observed variables with a small number of abstract latent factors.

The technique is very similar to PCA, but there is a fundamental difference. PCA aims at finding a series of rotations in such a way that the first axis corresponds with the direction of most variance, and each subsequent orthogonal axis explains the most of the remaining variance. In other words, PCA does not fit an explicit model. FA, on the other hand, does. For a mean-centered matrix X, the FA model is

$$X = LF + U \tag{4.7}$$

where L is the matrix of loadings on the *common factors* F, and U is a matrix of *specific factors*, also called *uniquenesses*. The common factors again are linear combinations of the original variables, and the scores present the positions of the samples in the new coordinate system. The result is a set of latent factors that capture as much variation, shared between variables, as possible. Variation that is unique to one specific variable will end up in the specific factors. Especially in fields like psychology it is customary to try and interpret the common factors like in the original approach by Spearman. Summarizing, it can be stated that PCA tries to represent as much as possible of the diagonal elements of the covariance matrix, whereas FA aims at reproducing the off-diagonal elements (Jackson 1991).

There is considerable confusion between PCA and FA, and many examples can be found where PCA models are actually called factor analysis models: one reason is that the simplest way (in a first approximation) to estimate the FA model of Equation 4.7 is to perform a PCA – this method of estimation is called Principal Factor Analysis. However, other methods exist, e.g., based on Maximum Likelihood, that provide more accurate models. The second source of confusion is that for spectroscopic data in particular, scientists are often trying to interpret the PCs of PCA. In that sense, they are more interested in the FA model than in the model-free transformation given by PCA.

Factor analysis is available in R as the function factanal, from the **stats** package (loaded by default). Application to the wine data is straightforward:

```
> (wines.fa <- factanal(wines.sc, 3, scores = "regression"))

Call:
factanal(x = wines.sc, factors = 3, scores = "regression")

Uniquenesses:
          alcohol        malic acid              ash
            0.393             0.729            0.524
    ash alkalinity         magnesium     tot. phenols
            0.068             0.843            0.198
       flavonoids non-flav. phenols          proanth
            0.070             0.660            0.559
         col. int.          col. hue         OD ratio
            0.243             0.503            0.248
          proline
            0.388


Loadings:
                    Factor1 Factor2 Factor3
alcohol                      0.776
malic acid          -0.467            0.211
ash                          0.287   0.626
ash alkalinity      -0.297  -0.313   0.864
magnesium            0.119   0.367
tot. phenols         0.825   0.346
flavonoids           0.928   0.262
non-flav. phenols   -0.533  -0.140   0.192
proanth              0.621   0.226
col. int.           -0.412   0.751   0.153
col. hue             0.653  -0.206  -0.170
OD ratio             0.865
proline              0.355   0.684  -0.134

                    Factor1 Factor2 Factor3
SS loadings           4.005   2.261   1.310
Proportion Var        0.308   0.174   0.101
Cumulative Var        0.308   0.482   0.583

Test of the hypothesis that 3 factors are sufficient.
The chi square statistic is 159.14 on 42 degrees of freedom.
The p-value is 1.57e-15
```

The default print method for a factor analysis object shows the uniquenesses, i.e., those parts that cannot be explained by linear combinations of other variables. The uniquenesses of the variables ash alkalinity and flavonoids, e.g., are very low, indicating that they may be explained well by the other variables. The loadings are printed in such a way as to draw attention to patterns: only three digits are shown after the decimal point, and smaller loadings are not printed.

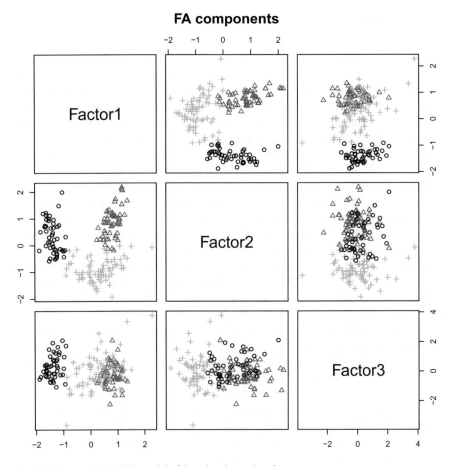

Fig. 4.11 Scores for the FA model of the wine data using three components

Scores can be calculated in several different ways, indicated by the `scores` argument of the `factanal` function. Differences between the methods are usually not very large; consult the manual pages of `factanal` to get more information about the exact implementation. The result for the regression scores, shown in Fig. 4.11, is only slightly different from what we have seen earlier with PCA; the familiar horse shoe is again visible in the first two components.

In Factor Analysis it is usual to rotate the components in such a way that the interpretability is enhanced. One of the ways to do this is to require that as few loadings as possible have large values, something that can be achieved by a so-called *varimax* rotation. This rotation is applied by default in `factanal`, and is the reason why the horseshoe is rotated compared to the PCA score plot in Fig. 4.1.

4.6.4   Discussion

Although there are some publications where ICA and FA are applied to data sets in the life sciences, their number is limited, and certainly much lower than the number of applications of PCA. There are several of reasons for this. Both ICA and FA do not have analytical solutions and require optimization to achieve their objectives, which takes more computing time, and can lead to different results, depending on the optimization settings. Moreover, several algorithms are available, each having slightly different definitions, which makes the results harder to compare and to interpret.

PCA, on the other hand, always gives the same result (if we disregard the sign ambiguity) when presented with a data set that is scaled in a certain way. In particular, PCA scores and loadings do not change when the number of component is changed. Since choosing the "right" number of components can be quite difficult, this is a real advantage over applying ICA, where earlier components depend on the total number of components taken into account. In a typical application, there are so many choices to make with respect to preprocessing, scaling, outlier detection and others, that there is a healthy tendency to choose methods that have as few tweaking possibilities as possible—if not, one can spend forever investigating the effects of small differences in analysis parameters. Nevertheless, there are cases where ICA, FA, or other dimension reduction methods can have definite advantages over PCA, and it can pay to check that.

The main aim of PCA is dimension reduction, often for visualization purposes, but scientists in the last decades have been remarkably creative in finding new and intriguing applications of the method. Very often, "classical" methods that cannot be applied to wide data matrices suddenly become possible after a PCA (see the chapters on regression and classification, for example). Another general use of PCA is as a form of denoising: one then assumes that the (majority of) the signal is contained in the first PCs and that later PCs only (or mostly) contain noise. In the latter application, the most important thing is not to take too few PCs, and usually one is quite generous in selecting the cut-off. Section 11.2 shows an important application, where a variant of PCA is used to identify outliers in multivariate data sets.

Chapter 5
Self-Organizing Maps

In PCA, the most outlying data points determine the direction of the PCs—these are the ones contributing most to the variance. This often results in score plots showing a large group of points close to the center. As a result, any local structure is hard to recognize, even when zooming in: such points are not important in the determination of the PCs. One approach is to select the rows of the data matrix corresponding to these points, and to perform a separate PCA on them. Apart from the obvious difficulties in deciding which points to leave out and which to include, this leads to a cumbersome and hard to interpret two-step approach. It would be better if a projection can be found that *does* show structure, even within very similar groups of points.

Self-organizing maps (SOMs, Kohonen 2001), sometimes also referred to as Kohonen maps after their inventor, Teuvo Kohonen, offer such a view. Rather than providing a continuous projection into $\mathbb{R}^2$, SOMs map all data to a set of discrete locations, organized in a regular grid. Associated with every location is a prototypical object, called a *codebook vector*. This usually does not correspond to any particular object, but rather represents part of the space of the data. The complete set of codebook vectors therefore can be viewed as a concise summary of the original data. Individual objects from the data set can be mapped to the set of positions, by assigning them to the unit with the most similar codebook vectors.

The effect is shown in Fig. 5.1. A two-dimensional point cloud is simulated where most points are very close to the origin.[1] The codebook vectors of a 5-by-5 rectangular SOM are shown in black; neighboring units in the horizontal and vertical directions are connected by lines. Clearly, the density of the codebook vectors is greatest in areas where the density of points is greatest. When the codebook vectors are shown at their SOM positions the plot on the right in Fig. 5.1 emerges, where individual objects are shown at a random position close to "their" codebook vector. The codebook vectors in the middle of the map are the ones that cover the center of the data density, and

---

[1] The point cloud is a superposition of two bivariate normal distributions, centered at the origin and with diagonal covariance matrices. The first has unit variance and contains 100 points; the other, containing 500 points, has variances of 0.025.

© Springer-Verlag GmbH Germany, part of Springer Nature 2020
R. Wehrens, *Chemometrics with R*, Use R!,
https://doi.org/10.1007/978-3-662-62027-4_5

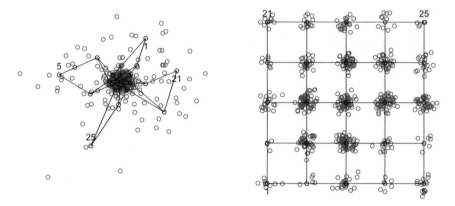

Fig. 5.1 Application of a 5-by-5 rectangular SOM to 600 bivariate normal data points. Left plot: location of codebook vectors in the original space. Right plot: location of data points in the SOM

one can see that these contain most data points. That is, relations within this densely populated area can be investigated in more detail.

5.1   Training SOMs

A SOM is trained by repeatedly presenting the individual samples to the map. At each iteration, the current sample is compared to the codebook vectors. The most similar codebook vector (the "winning unit") is then shifted slightly in the direction of the mapped object. This is achieved by replacing it with a weighted average of the old values of the codebook vector, cv_i, and the values of the new object obj:

$$cv_{i+1} = (1 - \alpha)\, cv_i + \alpha\, obj \tag{5.1}$$

The weight, also called the learning rate α, is a small value, typically in the order of 0.05, and decreases during training so that the final adjustments are very small.

As we shall see in Sect. 6.2.1, the algorithm is very similar in spirit to the one used in k-means clustering, where cluster centers and memberships are alternatingly estimated in an iterative fashion. The crucial difference is that not only the winning unit is updated, but also the other units in the "neighborhood" of the winning unit. Initially, the neighborhood is fairly large, but during training it decreases so that finally only the winning unit is updated. The effect is that neighboring units in general are more similar than units far away. Or, to put it differently, moving through the map by jumping from one unit to its neighbor would see gradual and more or less smooth transitions in the values of the codebook vectors. This is clearly visible in the mapping of the autoscaled wine data to a 5-by-4 SOM, using the **kohonen** package:

Fig. 5.2 Codebook vectors
for a SOM mapping of the
autoscaled wine data. The
thirteen variables are shown
counterclockwise, beginning
in the first quadrant

Codes plot

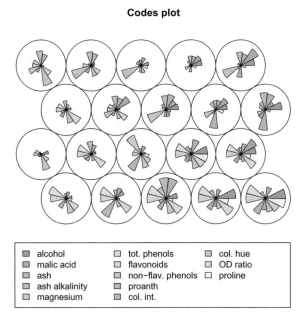

■ alcohol	▢ tot. phenols	■ col. hue
■ malic acid	▢ flavonoids	▢ OD ratio
■ ash	■ non-flav. phenols	▢ proline
■ ash alkalinity	■ proanth	
▢ magnesium	■ col. int.	

```
> wines.som <- som(wines.sc, somgrid(5, 4, "hexagonal"))
> plot(wines.som, type = "codes")
```

The result is shown in Fig. 5.2. Units in this example are arranged in a hexagonal
fashion and are numbered row-wise from left to right, starting from the bottom left.
The first unit at the bottom left for instance, is characterized by relatively large values
of alcohol, flavonoids and proanth; the second unit, to the right of the first,
has lower values for these variables, but still is quite similar to unit number one.

The codebook vectors are usually initialized by a random set of objects from the
data, but also random values in the range of the data can be employed. Sometimes
a grid is used, based on the plane formed by the first two PCs. In practice, the
initialization method will hardly ever matter; however, starting from other random
initial values will lead to different maps. The conclusions drawn from the different
maps, however, tend to be very similar.

The training algorithm for SOMs can be tweaked in many different ways. One
can, e.g., update units using smaller changes for units that are further away from
the winning unit, rather than using a constant learning rate within the neighborhood.
One can experiment with different rates of decreasing values for learning rate and
neighborhood size. One can use different distance measures. Regarding topology,
hexagonal or rectangular ordering of the units is usually applied; in the first case, each
unit has six equivalent neighbors, unless it is at the border of the map, in the second
case, depending on the implementation, there are four or eight equivalent neighbors.
The most important parameter, however, is the size of the map. Larger maps allow

for more detail, but may contain more empty units as well. In addition, they take more time to be trained. Smaller maps are more easy to interpret; groups of units with similar characteristics are more easily identified. However, they may lack the flexibility to show specific groupings or structure in the data. Some experimentation usually is needed. As a rule of thumb, one can consider the object-to-unit ratio, which can lead to useful starting points. In image segmentation applications, for instance, where hundreds of thousands of (multivariate) pixels need to be mapped, one can choose a map size corresponding to an average of several hundreds of pixels per unit; in other applications where the number of samples is much lower, a useful object-to-unit ratio may be five. One more consideration may be the presence of class structure: for every class, several units should be allocated. This allows intra-class structure to be taken into account, and will lead to a better mapping.

Finally, there is the option to close the map, i.e., to connect the left and right sides of the map, as well as the bottom and top sides. This leads to a toroidal map, resembling the surface of a closed tube. In such a map, all differences between units have been eliminated: there are no more edge units, and they all have the same number of neighbors. Whereas this may seem a desirable property, there are a number of disadvantages. First, it will almost certainly be depicted as a regular map with edges, and when looking at the map one has to remember that the edges in reality do not exist. In such a case, similar objects may be found in seemingly different parts of the map that are, in fact, close together. Another pressing argument against toroidal maps is that in many cases the edges serve a useful purpose: they provide refuge for objects that are quite different from the others. Indeed, the corners of non-toroidal maps often contain the most distinct classes.

5.2   Visualization

Several different visualization methods are provided in the **kohonen** package: one can look at the codebook vectors, the mapping of the samples, and one can also use SOMs for prediction. Here, only a few examples are shown. For more information, consult the manual pages of the plot.kohonen function, or the software description (Wehrens and Buydens 2007; Wehrens and Kruisselbrink 2018).

For multivariate data, the locations of the codebook vectors can not be visualized as was done for the two-dimensional data in Fig. 5.1. In the **kohonen** package, the default is to show segment plots, such as in Fig. 5.2 if the number of variables is smaller than 15, and a line plot otherwise. One can also zoom in and concentrate on the values of just one of the variables:

```
> for (i in c(1, 8, 11, 13))
+   plot(wines.som, "property",
+        property = getCodes(wines.som, 1)[, i],
+        main = colnames(wines)[i])
```

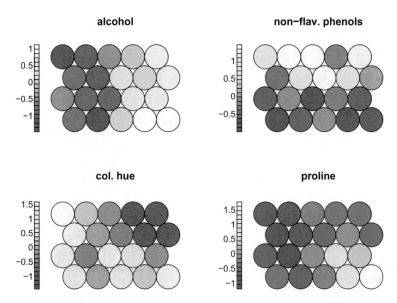

Fig. 5.3 Separate maps for the contributions of individual variables to the codebook vectors of the SOM shown in Fig. 5.2

Clearly, in these plots, shown in Fig. 5.3, there are regions in the map where specific variables have high values, and other regions where they are low. Areas of high values and low values are much more easily recognized than in Fig. 5.2. Note the use of the accessor function `getCodes` here.

Perhaps the most important visualization is to show which objects map in which units. In the **kohonen** package, this is achieved by supplying the the `type` = `"mapping"` argument to the plotting function. It allows for using different plotting characters and colors (see Fig. 5.4):

```
> plot(wines.som, type = "mapping",
+       col = as.integer(vintages), pch = as.integer(vintages))
```

Again, one can see that the wines are well separated. Some class overlap remains, especially for the Grignolinos (pluses in the figure). These plots can be used to make predictions for new data points: when the majority of the objects in a unit are, e.g., of the Barbera class, one can hypothesize that this is also the most probably class for future wines that end up in that unit. Such predictions can play a role in determining authenticity, an economically very important application.

Since SOMs are often used to detect grouping in the data, it makes sense to look at the codebook vectors more closely, and investigate if there are obvious class boundaries in the map—areas where the differences between neighboring units are relatively large. Using a color code based on the average distance to neighbors one can get a quick and simple idea of where the class boundaries can be found. This

Fig. 5.4 Mapping of the 177
wine samples to the SOM
from Fig. 5.2. Circles
correspond to Barbera,
triangles to Barolo, and
pluses to Grignolino wines

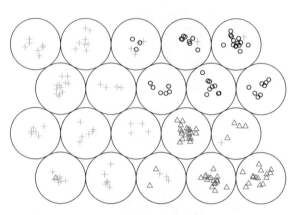

Fig. 5.5 Summed distances
to direct neighbors: the
U-matrix plot for the
mapping of the wine data

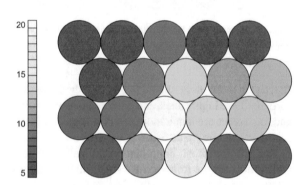

idea is often referred to as the "U-matrix" (Ultsch 1993), and can be employed by
issuing:

```
> plot(wines.som, type = "dist.neighb")
```

The resulting plot is shown in Fig. 5.5. The map is too small to really be able to see
class boundaries, but one can see that the centers of the classes (the bottom left corner
for Barbera, the bottom right corner for Barolo, and the top row for the Grignolino
variety) correspond to areas of relatively small distances, i.e., high homogeneity.

Training progress, and an indication of the quality of the mapping, can be obtained
using the following plotting commands:

```
> par(mfrow = c(1, 2))
> plot(wines.som, "changes")
> plot(wines.som, "quality")
```

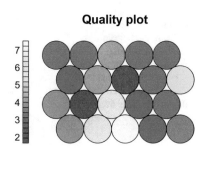

Fig. 5.6 Quality parameters for SOMs: the plot on the left shows the decrease in distance between objects and their closest codebook vectors during training. The plot on the right shows the mean distances between objects and codebook vectors per unit

This leads to the plots in Fig. 5.6. The left plot shows the average distance (expressed per variable) to the winning unit during the training iterations, and the right plot shows the average distance of the samples and their corresponding codebook vectors after training. Note that the latter plot concentrates on distances *within* the unit whereas the U-matrix plot in Fig. 5.5 visualizes average distances *between* neighboring units.

Finally, an indication of the quality of the map is given by the mean distances of objects to their units:

```
> summary(wines.som)
SOM of size 5x4 with a hexagonal topology
  and a bubble neighbourhood function.
The number of data layers is 1.
Distance measure(s) used: sumofsquares.
Training data included: 177 objects.
Mean distance to the closest unit in the map: 3.646.
```

The summary function indicates that an object, on average, has a distance of 3.6 units to its closest codebook vector. The plot on the left in Fig. 5.6 shows that the average distance drops during training: codebook vectors become more similar to the units that are mapped to them. The plot on the right, finally, shows that the distances within units can be quite different. Interestingly, some of the units with the largest spread only contain Grignolinos (units 2 and 8), so the variation can not be attributed to class overlap alone.

Fig. 5.7 Mapping of the prostate data. The cancer samples (pca) lie in a broad band from the bottom right to the top of the map. Control samples are split in two groups on either side of the pca samples. There is considerable class overlap

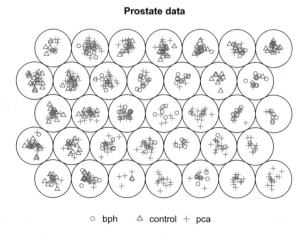

Prostate data

○ bph   △ control   + pca

5.3   Application

The main attraction of SOMs lies in the applicability to large data sets; even if the data are too large to be loaded in memory in one go, one can train the map sequentially on (random) subsets of the data. It is also possible to update the map when new data points become available. In this way, SOMs provide a intuitive and simple visualization of large data sets in a way that is complementary to PCA. An especially interesting feature is that these maps can show grouping of the data without explicitly performing a clustering. In large maps, sudden transitions between units, as visualized by, e.g., a U-matrix plot, enable one to view the major structure at a glance. In smaller maps, this often does not show clear differences between groups—see Fig. 5.5 for an example. One way to find groups is to perform a clustering of the individual codebook vectors. The advantage of clustering the codebook vectors rather than the original data is that the number of units is usually orders of magnitude smaller than the number of objects.

As a practical example, consider the mapping of the 654 samples from the prostate data using the complete, 10,523-dimensional mass spectra in a 7-by-5 map. This would on average lead to almost twenty samples per unit and, given the fact that there are three classes, leave enough flexibility to show within-class structure as well:

```
> X <- t(Prostate2000Raw$intensity)
> prostate.som <- som(X, somgrid(7, 5, "hexagonal"))
```

The plot in Fig. 5.7 is produced with the following code:

```
> types <- as.integer(Prostate2000Raw$type)
> trellis.cols <- trellis.par.get("superpose.symbol")$col[c(2, 3, 1)]
> plot(prostate.som, "mapping", col = trellis.cols[types],
+      pch = types, main = "Prostate data")
```

```
> legend("bottom", legend = levels(Prostate2000Raw$type),
+        col = trellis.cols, pch = 1:3, ncol = 3, bty = "n")
```

Clearly, there is considerable class overlap, as may be expected when calculating distances over more than 10,000 variables. Some separation can be observed, however, especially between the cancer and control samples. To investigate differences between the individual units, one can plot the codebook vectors of some of the units containing (predominantly) objects from one class only, corresponding to the three right-most units in the plot in Fig. 5.7:

```
> units <- c(7, 21, 35)
> unitfs <- paste("Unit", units)
> prost.plotdf <-
+    data.frame(mz = Prostate2000Raw$mz,
+               intensity = c(t(getCodes(prostate.som, 1)[units, ])),
+               unit = rep(factor(unitfs, levels = unitfs),
+                          each = length(Prostate2000Raw$mz)))
> xyplot(intensity ~ mz | unit, data = prost.plotdf, type = "l",
+        scale = list(y = "free"), as.table = TRUE,
+        xlab = bquote(italic(.("m/z"))~.("(Da)")),
+        groups = unit, layout = c(1, 3),
+        panel = function(...) {
+          panel.abline(v = c(3300, 4000, 6000, 6200),
+                       col = "gray", lty = 2)
+          panel.xyplot(...)
+        })
```

These codebook vectors, shown in Fig. 5.8, display appreciable differences. The cancer samples from Unit 7, for instance, are missing the large peaks at 3,100 and 4,000 Da that are present in the other two units but contain very clear signals around 6,100 and 6,200 Da, where the others have nothing.

5.4 R Packages for SOMs

The **kohonen** package used in this chapter, originally based on the **class** package (Venables and Ripley 2002), has several noteworthy features not discussed yet (Wehrens and Kruisselbrink 2018). It can use distance functions other than the usual Euclidean distance, which might be extremely useful for some data sets, often avoiding the need for prior data transformations. One example is the WCC function mentioned earlier: this can be used to group sets of X-ray powder diffractograms where the position rather than the position of peaks contains the primary information (Wehrens and Willighagen 2006; Wehrens and Kruisselbrink 2018). For numerical variables, the sum-of-squares distance is the default (slightly faster than the Euclidean distance); for factors, the Tanimoto distance. In the **kohonen** package

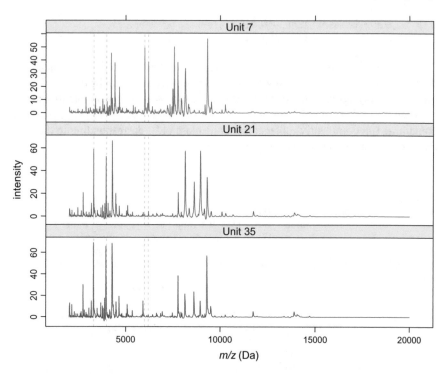

Fig. 5.8 Codebook vectors for three units from the far-right side of the map in Fig. 5.7, containing only samples from one class: unit 7 contains pca samples, unit 21 mostly bph samples and unit 35 control samples. Vertical gray lines indicate mass-to-charge ratios mentioned in the text

it is possible to supply several different data layers, where the rows in each layer correspond to different bits of information on the same objects. Separate distance functions can be defined for each single layer, which are then combined into one overall distance measure using weights that can be defined by the user. Apart from the usual "online" training algorithm described in this chapter, a "batch" algorithm is implemented as well, where codebook vectors are not updated until all records have been presented to the map. One advantage of the batch algorithm is that it dispenses with one of the parameters of the SOM: the learning rate α is no longer needed. The main disadvantage is that it is sometimes less stable and more likely to end up in a local optimum. The batch algorithm also allows for parallel execution by distributing the comparisons of objects to all codebook vectors over several cores (Lawrence et al. 1999) which may lead to considerable savings with larger data sets (Wehrens and Kruisselbrink 2018).

Several other packages are available from repositories like CRAN. One example is the **som** package (Yan 2016). This package implements the online and batch algorithms and provides great flexibility in setting training parameters. The **somoclu** package implements a general SOM toolbox supporting parallel computation, also on GPUs (Wittek et al. 2017). It uses the **kohonen** plotting functions for visualization. A package providing a **shiny** (Chang et al. 2018) web interface is **SOMbrero** (Olteanu and Villa-Vialaneix 2015). Here, one can use numerical data, contingency tables as well as distance matrices as primary input data. Finally, the Stuttgart Neural Network Simulator (**SNNS**) provides SOMs as one of many types of neural networks (Bergmeir and Benítez 2012).

5.5 Discussion

Conceptually, the SOM is most related to MDS, seen in Sect. 4.6.1. Both, in a way, aim to find a configuration in two-dimensional space that represents the distances between the samples in the data. Whereas metric forms of MDS focus on the preservation of the actual distances, SOMs provide a topological mapping, preserving the *order* of the distances, at least for the smallest ones. Because of this, an MDS mapping is often dominated by the larger distances, even when using methods like Sammon mapping, and the configuration of the finer structure in the data may not be well preserved. In SOMs, on the other hand, a big distance between the positions of two samples in the map does not mean that they are very dissimilar: if the map is too large, and not well trained, two regions in the map that are far apart may very well have quite similar codebook vectors. What one *can* say is that objects mapped to the same or to neighboring units are likely to be similar.

Both MDS and SOMs operate using distances rather than the original data to determine the position in the low-dimensional representation of the data. This can be a considerable advantage when working with high-dimensional data: even when the number of variables is in the tens or hundreds of thousands, the distances between objects can be calculated fairly quickly. Obviously, MDS, in particular, runs into trouble when the number of samples gets large—SOMs can handle that more easily because of the iterative training procedure employed. It is not even necessary to have all the data in memory simultaneously.

Using SOMs is doubtful when the number of samples is low, although applications have been published with fewer than fifty objects. If the number of units in the map is much smaller than the number of objects in such cases, one loses the advantage of the spatial smoothness in the map, and one could just as well perform a clustering; if the number of units approaches the number of objects, it is more likely than not that the majority of the objects will occupy a unit by itself, which is not very informative either.

One should realize that in the case of correlated variables the distances that are calculated may be a bit biased: a group of highly correlated variables will have a major influence on the distance between objects. In areas like, e.g., quantitative structure-activity relationships (QSAR), it is usual to calculate as many chemical structure descriptors as possible in order to define the two- or three-dimensional structure of a set of compounds. Many of these descriptors are variations on a theme: some groups measure properties related to dipole, polarizability, surface area, etcetera. The influence of one single descriptor capturing information that is unrelated to the hundreds of other descriptors can easily be lost when calculating distances. For SOMs, one simple solution is to decorrelate the (scaled) data, e.g., using PCA, and to calculate distances using the scores.

Chapter 6
Clustering

As we saw earlier in the visualizations provided by methods like PCA and SOM, it is often interesting to look for structure, or groupings, in the data. However, these methods do not explicitly define clusters; that is left to the pattern recognition capabilities of the scientist studying the plot. In many cases, however, it is useful to rely on somewhat more formal methods, and this is where clustering methods come in. They are usually based on object-wise similarities or distances, and since the late nineties have become hugely popular in the area of high-throughput measurement techniques in biology, such as DNA microarrays. There, the activities of tens of thousands of genes are measured, often as a function of a specific treatment, or as a time series. Of course, the question is which genes show the same activity pattern: if an unknown gene has much the same behavior as another gene of which it is known that it is involved in a process like cell differentiation, one can hypothesize that the unknown gene is somehow related to this process as well.

With only a slight exaggeration one could say that there are about as many clustering algorithms as there are scientists and by no means do these methods always give the same results. Modern software packages have made many of these clustering methods available to a wide audience; unfortunately, this provides the temptation to try all methods in order to get the result one is looking for, rather than the result that is suggested by the data. There are no formal rules to help you decide which clustering method to use.

One of the reasons for this is that most clustering methods are heuristic in nature, rather than that they stem from solid statistical foundations. Moreover, assessing the quality of the clustering, or *validation*, is a problem: since the "real" clustering is by definition unknown (otherwise it would be more appropriate to use a supervised approach such as the classification methods described in Chap. 7) we can not say that one clustering is better than the other. Also cluster characteristics (sphericity, density, ...) can not be used for this, since different clustering methods "optimize" different criteria. It is often difficult for users to get a good idea of the behavior of the separate methods, since our visualization abilities break down in more than three dimensions, and at the same time the assumptions behind the clustering methods are often unknown.

© Springer-Verlag GmbH Germany, part of Springer Nature 2020
R. Wehrens, *Chemometrics with R*, Use R!,
https://doi.org/10.1007/978-3-662-62027-4_6

In this chapter, we concentrate on several popular classes of methods. Hierarchical methods are represented by *single, average* and *complete linkage*, respectively, while *k-means* is an example of partitional methods. Both yield "crisp" clusterings; objects belong to exactly one cluster. More sophisticated methods lead to a clustering where membership values are assigned to each object: the object can be assigned to the cluster with the highest membership value. An example is given by model-based clustering methods.

6.1 Hierarchical Clustering

Quite often, data have a hierarchical structure in the sense that groups consist of mutually exclusive sub-groups. This is often visualized in a tree-like structure, called a dendrogram. The dendrogram presents an intuitive and appealing way for visualizing the hierarchical structure: the *y*-axis indicates the "distance" between different groups, whereas the connections show where successive splits (or joins) take place.

Hierarchical clustering starts with a square matrix containing distances or (dis)similarities; in the following we will assume we have the data in the form of distances. It is almost always performed in a bottom-up fashion. Starting with all objects in separate clusters, one looks for the two most similar clusters and joins them. Then, the distance matrix is updated. There are several possibilities to determine the distance between clusters. One option is to take the shortest distance between clusters. In Fig. 6.1 this would correspond to the distance between objects d and D. This choice leads to the *single-linkage* algorithm. It joins two groups if any members of both groups are close together, a strategy that is sometimes also referred to as friends-of-friends: "any friend of yours is my friend, too!".

The opposite strategy is *complete linkage* clustering: there, the distance between clusters is determined by the objects in the respective clusters that are furthest apart—in Fig. 6.1 objects a and A. In other words: to belong to the same cluster, the distances to *all* cluster members must be small.[1] This strategy leads to much more compact and

Fig. 6.1 Distances between clusters: single linkage, average linkage and complete linkage consider the closest points, the averages, and the farthest points, respectively

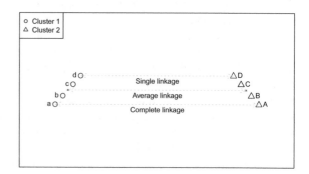

---

[1] We can only be friends if all our friends are friends of *both* of us.

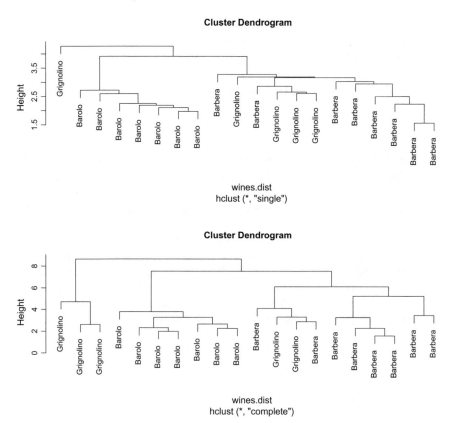

Fig. 6.2 Single linkage clustering (top) and complete linkage clustering (bottom) of 20 samples from the `wine` data

equal-sized clusters. Of course, intermediate strategies are possible, too. Taking the distance between cluster means leads to *average linkage*. *Ward's method* explicitly takes into account the cluster size in calculating a weighted average, and in many cases gives very similar result to average linkage.

Let us see how this works by clustering a random subset of the wine data. In R hierarchical clustering is available through function `hclust`, which takes an object of class `dist` as its first argument:

```
> subset <- sample(nrow(wines), 20)
> wines.dist <- dist(wines.sc[subset, ])
> wines.hcsingle <- hclust(wines.dist, method = "single")
> plot(wines.hcsingle, labels = vintages[subset])
```

This leads to the dendrogram at the top in Fig. 6.2. When we go down in distance, starting from the top, one Grignolino sample is split off from the main branch as a singleton before the whole Barolo cluster is identified. Going down even further,

individual Grignolino and Barbera samples are split off before arriving at a cluster
of Grignolino wines, and a cluster of Barberas.

Also the complete linkage dendrogram in the bottom panel of Fig. 6.2, suggesting a
four- or five-cluster solution, shows the confusion between Barberas and Grignolinos,
and separate, pure, Barolo and Grignolino clusters. This plot is obtained by:

```
> wines.hccomplete <- hclust(wines.dist, method = "complete")
> plot(wines.hccomplete, labels = vintages[subset])
```

The layout of the dendrogram is very different from the single-linkage one: there, the
typical friends-of-friends behaviour is observed, where single objects are gradually
added to one large group, in addition to a number of singletons. In complete linkage
one often finds more clear distinctions between groups of samples, as is the case
here.

In principle, a dendrogram from a hierarchical clustering method in itself is not
yet a clustering, since it does not give a grouping as such. However, these can be
obtained by "cutting" the diagram at a certain height: all objects that are connected are
supposed to be in one and the same cluster. For this, function cutree is available,
which either takes the height at which to cut, or the number of clusters to obtain as
an argument. In this case, let's cut at a height of 3:

```
> wines.cl.single <- cutree(wines.hcsingle, h = 3)
> table(wines.cl.single, vintages[subset])

wines.cl.single Barbera Barolo Grignolino
              1       0      7          0
              2       1      0          0
              3       1      0          3
              4       0      0          1
              5       1      0          0
              6       5      0          0
              7       0      0          1
```

The clustering is very good in the sense that there are almost no mixed clusters
containing samples from more than one type; on the other hand, the Barbera wines
are split over four different clusters. Cutting the dendrogram at a height larger than
three will lead to fewer clusters but inevitably also to more mixed clusters. Conversely,
lowering the height at which one cuts leads to more, and more pure clusters. What
is most useful needs to be determined on a case-to-case basis.

Now we turn to the complete data set, and recalculate the clusterings. Single
linkage, cut at a height to obtain three clusters, does not show anything useful:

```
> wines.dist <- dist(wines.sc)
> wines.hcsingle <- hclust(wines.dist, method = "single")
> table(vintages, cutree(wines.hcsingle, k = 3))
```

```
vintages        1  2  3
   Barbera     48  0  0
   Barolo      58  0  0
   Grignolino  67  3  1
```

Almost all samples are in cluster 1, and small bits of the data set (all Grignolino samples) are chipped off leading to clusters 2 and 3, each with only very few elements. On the other hand, the three-cluster solution from complete linkage is already quite good:

```
> wines.hccomplete <- hclust(wines.dist, method = "complete")
> table(vintages, cutree(wines.hccomplete, k = 3))
```

```
vintages        1  2  3
   Barbera      3  0 45
   Barolo      50  8  0
   Grignolino  14 52  5
```

Cluster 1 corresponds to mainly Barolo wines, cluster two to Grignolinos, and cluster three to the Barberas. Of course, there still is significant overlap between the clusters.

Hierarchical clustering methods enjoy great popularity: the intuitive visualization through dendrograms is one of the main reasons. These also provide the opportunity to see the effects of increasing the number of clusters, without actually recalculating the cluster structure. Obviously, hierarchical clustering will work best when the data actually have a hierarchical structure: that is, when clusters contain subclusters, or when some clusters are more similar than others. In practice, this is quite often the case.

A further advantage is that the clustering is unique: no random element is involved in creating the cluster model. For many other clustering methods, this is not the case. Note that the uniqueness property is present only in the case that there are no ties in the distances. If there are, one may obtain several different dendrograms, depending on the order of the data and the actual implementation of the software. Usually, the first available merge with the minimal distance is picked. When equal distances are present, one or more equivalent merges are possible, which may lead to different dendrograms. An easy way to investigate this is to repeat the clustering many times on distance matrices from data where the rows have been shuffled.

There are a number of drawbacks to hierarchical clustering, too. For data sets with many samples (more than ten thousand, say) these methods are less suitable. To start with, calculating the distance matrix may be very expensive, or even impossible. More importantly, interpreting the dendrograms quickly becomes cumbersome, and there is a real danger of over-interpretation. Examples where hierarchical methods are used with large data sets can be found in the field of DNA microarrays, where the ground-breaking paper of Eisen et al. (1998) seems to have set a trend.

There are a number of cases where the results of hierarchical clustering can be misleading. The first is the case where in reality there is no class structure. Cutting a dendrogram will always give you clusters: unfortunately, there is no warning light flashing when you investigate a data set with no class structure. Furthermore, even when there are clusters, they may be too close to separate, or they may overlap. In these cases it is impossible to conclude anything about individual cases (although it can still be possible to infer characteristics of the clusters as a whole). The two keys to get out of this conundrum are formed by the use of prior information, and by visualization. If you know class structure is present, and you already have information about part of that structure, the clustering methods that fail to reproduce that knowledge obviously are not performing well, and you are more likely to trust the results of the methods that do find what you already know. Another idea is to visualize the (original) data, and give every cluster a different color and plotting symbol. One can easily see if clusters are overlapping or are nicely separated. Note that the dendrogram can be visualized in a number of equivalent ways: the ordering of the groupings from left to right is arbitrary to some extent and may depend on your software package.

The **cluster** package in R also provides functions for hierarchical clustering: agnes implements single, average and complete linkage methods but also allows more control over the distance calculations using the method ="flexible" argument. In addition, it provides a coefficient measuring the amount of cluster structure, the "agglomerative coefficient", ac:

$$ac = \frac{1}{n} \sum_i (1 - m_i)$$

where the summation is over all n objects, and m_i is the ratio of the dissimilarity of the first cluster an object is merged to and the dissimilarity level of the final merge (after which only one cluster remains). Compare these numbers for three hierarchical clusterings of the wine data:

```
> wines.agness <- agnes(wines.dist, method = "single")
> wines.agnesa <- agnes(wines.dist, method = "average")
> wines.agnesc <- agnes(wines.dist, method = "complete")


> cbind(wines.agness$ac, wines.agnesa$ac, wines.agnesc$ac)
          [,1]    [,2]     [,3]
[1,] 0.53802 0.69945 0.81625
```

Complete linkage is doing the best job for these data, according to this quality measure.

6.2 Partitional Clustering

A completely different approach is taken by partitional clustering methods. Instead of starting with individual objects as clusters and progressively merging similar clusters, partitional methods choose a set of cluster centers in such a way that the overall distance of all objects to the closest cluster centers is minimised. Algorithms are iterative and usually start with random cluster centers, ending when no more changes in the cluster assignments of individual objects are observed. Again, many different flavours exist, each with its own characteristics. In general, however, these algorithms are very fast and are suited for large numbers of objects. The calculation of the complete distance matrix is unnecessary—only the distances to the cluster centers need to be calculated, where the number of clusters is much smaller than the number of objects—and this saves resources. Two examples will be treated here: k-means and k-medoids. The latter is a more robust version, where outlying observations do not influence the clustering to a large extent.

6.2.1 K-Means

The k-means algorithm is very simple and basically consists of two steps. It is initialized by a random choice of cluster centers, e.g., a random selection of objects in the data set or random values within the range for each variable. Then the following two steps are iterated:

1. Calculate the distance of an object to all cluster centers and assign the object to the closest center; do this for all objects.
2. Replace the cluster centers by the means of all objects assigned to them.

The quality of the final model can then be assessed by summing the distances of all objects to the centers of the clusters to which they are assigned. Note the similarity to the training of SOMs in Chap. 5, in particular to the batch training algorithm. The goals of the two methods, however, are quite different: SOMs aim at providing a suitable mapping to two dimensions, and the units should not be seen as individual clusters, whereas k-means explicitly focusses on finding a specific number of groups.

The basic R function is for k-means clustering conveniently called kmeans. Application to the wine data leads to the following result:

```
> (wines.km <- kmeans(wines.sc, centers = 3))
K-means clustering with 3 clusters of sizes 65, 51, 61

Cluster means:
     alcohol malic acid       ash ash alkalinity magnesium
1 -0.91833    -0.39533 -0.49050          0.16370 -0.483216
2  0.17364     0.86425  0.18718          0.51684 -0.064971
3  0.83336    -0.30131  0.36617         -0.60655  0.569222
    tot. phenols flavonoids non-flav. phenols   proanth col. int.
1     -0.071141   0.026589        -0.037096  0.065095  -0.89558
2     -0.971065  -1.206242         0.719152 -0.771710   0.93782
3      0.887680   0.980165        -0.561730  0.575837   0.17023
    col. hue OD ratio  proline
1   0.46141  0.28236 -0.74607
2  -1.15662 -1.28723 -0.40027
3   0.47535  0.77533  1.12965


Clustering vector:
   [1] 3 3 3 3 3 3 3 3 3 3 3 3 3 3 3 3 3 3 3 3 3 3 3 3 3 3 3 3 3 3
  [31] 3 3 3 3 3 3 3 3 3 3 3 3 3 3 3 3 3 3 3 3 3 3 3 3 3 3 3 3 1 1
  [61] 2 1 1 1 1 1 1 1 1 1 1 1 3 1 1 1 1 1 1 1 1 1 2 1 1 1 1 1 1 1
  [91] 1 1 1 1 3 1 1 1 1 1 1 1 1 1 1 1 1 1 1 1 1 1 1 1 1 1 1 2 1 1
 [121] 3 1 1 1 1 1 1 1 1 2 2 2 2 2 2 2 2 2 2 2 2 2 2 2 2 2 2 2 2 2
 [151] 2 2 2 2 2 2 2 2 2 2 2 2 2 2 2 2 2 2 2 2 2 2 2 2 2 2 2

Within cluster sum of squares by cluster:
[1] 559.30 326.35 382.19
 (between_SS / total_SS =  44.6 %)

Available components:

[1] "cluster"       "centers"      "totss"      "withinss"
[5] "tot.withinss"  "betweenss"    "size"       "iter"
[9] "ifault"
```

The algorithm not only returns the clustering of the individual objects, but also cluster-specific information such as the sum of squares, the cumulative distance of all cluster objects to the center of the cluster.

So the question now is: how good is the agreement with the vintages? Let's see:

```
> table(vintages, wines.km$cluster)

vintages      1  2  3
  Barbera     0 48  0
  Barolo      0  0 58
  Grignolino 65  3  3
```

Only six of the Grignolino samples are classified as Barbera and Barolo wines; a lot better than the complete-linkage solution.

The k-means algorithm enjoys great popularity through its simplicity, ease of interpretation, and speed. It does have a few drawbacks, however. We already mentioned the fact that one should pick the number of clusters in advance. In general, the correct number (if such a thing exists at all) is never known, and one will probably try several different clusterings with different numbers of clusters. Whereas hierarchical clustering delivers this in one go—the dendrogram only has to be cut at different positions—for k-means clustering (and partitional methods in general) one should repeat the whole clustering procedure. As already said, the results with four or five clusters may differ dramatically.

Worse, even a repeated clustering with the *same* number of clusters will give a different result, sometimes even a very different result. Remember that we start from a random initialization: an unlucky starting point may get the algorithm stuck in a local minimum. Repeated application, starting from different initial guesses, gives some idea of the variability. The kmeans function returns the within-cluster sums of squares for the separate clusters, which can be used as a quality criterion:

```
> wines.km <- kmeans(wines.sc, centers = 3)
> best <- wines.km
> for (i in 1:100) {
+    tmp <- kmeans(wines.sc, centers = 3)
+    if (sum(tmp$withinss) < sum(best$withinss))
+      best <- tmp
+ }
```

One can then pick the one that leads to the best description of the data or, equivalently, the smallest overall distance. In this particular case, the overall best solution is found every time—the wine data do not present that much of a problem. The kmeans function has a built-in argument for repeating the clustering and only returning the best solution. Thus, the loop in the previous example can be replaced by

```
> wines.km <- kmeans(wines.sc, centers = 3, nstart = 100)
```

Several minima with comparable overall distance measure may exist, so that different but equally good clustering solutions can be found by the algorithm.

6.2.2 K-Medoids

In k-means, cluster centers are given by the mean coordinates of the objects in that cluster. Since averages are very sensitive to outlying observations, the clustering may be dominated by a few objects, and the interpretation may be difficult. One way to resolve this is to assess clusterings with more groups than expected: the outliers may end up in a cluster of their own. A more practical alternative would be to use a more robust algorithm where the influence of outliers is diminished. One example is the k-medoids algorithm (Kaufman and Rousseeuw 1990), available in R through the function pam—Partitioning Around Medoids— in the **cluster** package. Rather than finding cluster centers at optimal positions, k-medoids aims at finding *k* representative objects within the data set. Typically, the sum of the distances is minimized rather than the sum of the squared distances, decreasing the importance of large distances.

Applied to the wine data, k-medoids gives the following result:

```
> (wines.pam <- pam(wines.dist, k = 3))
Medoids:
       ID
[1,]   35   35
[2,] 106  106
[3,] 148  148
Clustering vector:
  [1] 1 1 1 1 1 1 1 1 1 1 1 1 1 1 1 1 1 1 1 1 1 1 1 1 1 1 1 1 1 1
 [31] 1 1 1 1 1 1 1 1 1 1 1 1 1 1 1 1 1 1 1 1 1 1 1 1 1 1 1 1 2 2
 [61] 2 2 1 2 1 2 2 2 1 2 1 2 1 1 2 2 2 1 1 2 2 2 3 2 2 2 2 2 2 2
 [91] 2 2 2 2 1 1 2 1 2 2 2 2 2 2 2 2 2 2 1 1 2 2 2 2 2 2 2 2 2 1
[121] 1 2 2 1 2 2 2 2 2 3 3 3 3 3 3 3 3 3 3 3 3 3 3 3 3 3 3 3 3 3
[151] 3 3 3 3 3 3 3 3 3 3 3 3 3 3 3 3 3 3 3 3 3 3 3 3 3 3 3 3
Objective function:
 build   swap
2.9085 2.8086

Available components:
[1] "medoids"    "id.med"      "clustering" "objective"
[5] "isolation"  "clusinfo"    "silinfo"    "diss"
[9] "call"
```

The result presents the medoids with their row numbers. The objective function, the sum of the distances of objects to the medoids, is reported in two stages: the first stage serves to find a good initial set of medoids, whereas the second stage performs a local search, trying all possible medoid swaps until no more improvement can be found. In this rather simple example, the average distance after the second stage has decreased by 0.1, compared to the distances to the initial set of medoids—not a huge decrease.

The implementation of pam in the **cluster** package comes with additional visualization methods. The first is the "silhouette" plot (Kaufman and Rousseeuw 1990). It shows a quality measure for individual clusterings: object with a high silhouette width (close to 1) are very well clustered, while objects with low values lie in between two or more clusters. Objects with a negative value may be even in the wrong cluster. The silhouette width s_i of object i is given by:

$$s_i = \frac{b_i - a_i}{\max(a_i, b_i)}$$

where a_i is the average distance of object i to all other objects in the same cluster, and b_i is the smallest distance of object i to another cluster. Thus, the maximal value will be obtained in those cases where the intra-cluster distance a is much smaller than the inter-cluster distance b.

For the wine data clustering, the silhouette plot shown in Fig. 6.3 is obtained by:

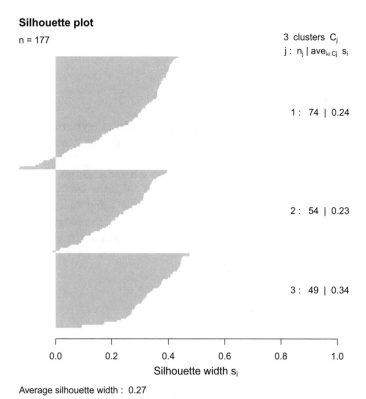

Fig. 6.3 Silhouette plot for the k-medoids clustering of the wine data. The three clusters contain 74, 54 and 49 objects, and have average silhouette widths of 0.24, 0.23 and 0.34, respectively

```
> plot(wines.pam, main = "Silhouette plot")
```

An overall measure of clustering quality can be obtained by averaging all silhouette widths. This is an easy way to decide on the most appropriate number of clusters:

```
> best.pam <- pam(wines.dist, k = 2)
> for (i in 3:10) {
+    tmp.pam <- pam(wines.dist, k = i)
+    if (tmp.pam$silinfo$avg.width < best.pam$silinfo$avg.width)
+        best.pam <- tmp.pam
+ }
> best.pam$medoids
[1]   12   56   34   97   91 163 125 148
```

In this case, eight clusters seem to give the clustering with the least ambiguity. The agreement with the true class labels is quite good:

```
> table(vintages, best.pam$clustering)

vintages        1   2   3   4   5   6   7   8
   Barbera      0   0   0   0   0  18   0  30
   Barolo      21  17  20   0   0   0   0   0
   Grignolino   0   1   5  20  15   5  24   1
```

Clusters 1, 2 and 3 correspond to the Barolo wines, and clusters 6 and 8 to the Barbera. Again, the Grignolino wines are the most difficult to cluster, and 12 Grignolino samples end up in clusters dominated by other wines.

For large data sets, pam is too slow; in the **cluster** package, an alternative is provided in the function clara (Kaufman and Rousseeuw 1990) which considers subsets of size sampsize. Each subset is partitioned using the same algorithm as in pam. The sets of medoids that result are used to cluster the complete data set, and the best set of medoids, i.e., the one for which the sum of the distances is minimal, is retained.

6.3 Probabilistic Clustering

In probabilistic clustering, sometimes also called fuzzy clustering, objects are not allocated to one cluster only. Rather, cluster memberships are used to indicate which of the clusters is more likely. If a "crisp" clustering result is needed, an object is assigned to the cluster with the highest membership value.

The most well-established methods are found in the area of mixture modelling, where individual clusters are represented by mixtures of parametric distributions, and the overall clustering is a weighted sum of the individual components (McLachlan and Peel 2000; Fraley and Raftery 2002). Usually, multivariate normal distributions are applied. In that case, assuming G clusters, the likelihood is given by

$$L(\tau, \mu, \Sigma | \mathbf{x}) = \prod_{i=1}^{n} \sum_{k=1}^{G} \tau_k \phi_k(\mathbf{x}_i | \mu_k, \Sigma_k) ,$$

where τ_k is the fraction of objects in cluster k, μ_k and Σ_k correspond to the cluster means and covariance matrices of cluster k, respectively, and ϕ_k is the density of cluster k. If the cluster labels would be known, one could estimate the unknown parameters τ_k, μ_k and Σ_k by maximizing the likelihood (for example). Vice versa, when these parameters are known, it is easy to calculate the conditional probabilities of belonging to class k:

$$z_{ik} = \phi_k(\mathbf{x}_i | \theta_k) / \sum_{j=1}^{K} \phi_j(\mathbf{x}_i | \theta_j)$$

These two steps are the components in the Expectation-Maximization algorithm (EM) (Dempster et al. 1977; McLachlan and Krishnan 1997): estimating the conditional probabilities is indicated with the E-step, whereas estimating the parameters (class means and variances, and mixing proportions) is the M-step. The conditional probabilities z_{ik} can also be seen as indicators of uncertainty: the larger $z_{i,max}$, the maximal value of all z_{ik} values for object i, the more certain the classification.

One can use the likelihood to determine what number of clusters is optimal. Of course, the likelihood will increase with the number of clusters, so one usually defines a penalty depending on the number of parameters that are estimated. Two popular measures are Akaike's Information Criterion (AIC) and the Bayesian Information Criterion (BIC). The AIC criterion (Akaike 1974) is defined by

$$AIC = -2 \log L + 2p \qquad (6.1)$$

where L is the likelihood and p the number of parameters in the model (here τ, μ, and Σ). The closely related BIC criterion (Schwarz 1978) uses a penalty that is usually stronger than the AIC penalty:

$$BIC = -2 \log L + p \log n \qquad (6.2)$$

The optimal model has a minimal value for AIC and/or BIC[2]—because of the more heavy penalty, BIC is likely to select slightly more parsimonious models than AIC. Several other criteria exist (McLachlan and Peel 2000). None of these is able to correctly identify the number of clusters in all cases, but in practice, differences are not very big and both AIC and BIC criteria are often used.

Several packages in R implement this form of clustering. In the **mclust** package (Fraley and Raftery 2003), for example, one can calculate BIC values for different numbers of clusters easily:

```
> wines.BIC <- mclustBIC(wines.sc, modelNames = "VVV")
> plot(wines.BIC)
```

This produces the plot in Fig. 6.4. The BIC value, here given in its negative form, has a maximum at two clusters, which will be the model picked by the function mclustModel if no specific number of clusters is given. Alternatively, one can specify a specific number of clusters by providing a value for the G argument:

```
> wines.mclust2 <- mclustModel(wines.sc, wines.BIC)
> wines.mclust3 <- mclustModel(wines.sc, wines.BIC, G = 3)
```

One can make scatter plots at specific combinations of variables with the coordProj function, visualizing the clustering in low-dimensional subspaces. Also the uncertainties, given by $1 - z_{i,max}$, can be visualized. This provides an easy way to compare the two- and three-cluster solutions graphically:

---

[2]Especially for the BIC value, one often sees the negative form so that maximization will lead to an optimal model. This is also the definition by (Schwarz 1978).

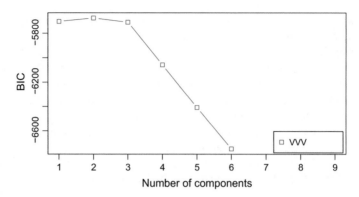

Fig. 6.4 BIC values for clustering the autoscaled wine data with **mclust**. The label "VVV" indicates a completely unconstrained model. The optimal model has two clusters

```
> par(mfrow = c(2, 2))
> coordProj(wines.sc, dimens = c(7, 13),
+           parameters = wines.mclust2$parameters,
+           z = wines.mclust2$z, what = "classification")
> title("2 clusters: classification")
> coordProj(wines.sc, dimens = c(7, 13),
+           parameters = wines.mclust3$parameters,
+           z = wines.mclust3$z, what = "classification")
> title("3 clusters: classification")
> coordProj(wines.sc, dimens = c(7, 13),
+           parameters = wines.mclust2$parameters,
+           z = wines.mclust2$z, what = "uncertainty")
> title("2 clusters: uncertainty")
> coordProj(wines.sc, dimens = c(7, 13),
+           parameters = wines.mclust3$parameters,
+           z = wines.mclust3$z, what = "uncertainty")
> title("3 clusters: uncertainty")
```

The result, here for the variables `flavonoids` and `proline`, is shown in Fig. 6.5. The top row shows the classifications of the two- and three-cluster models, respectively. The bottom row shows the corresponding uncertainties.

Just like with k-means and k-medoids, the clustering using the EM algorithm needs to be kick-started with an initial guess. This may be a random initialization, but the EM algorithm has a reputation for being slow to converge, and an unlucky guess may lead into a local optimum. In `mclust`, the initialization is done by hierarchical clustering.[3] This has the advantage that initial models for many different numbers of clusters can be generated quickly. Moreover, this initialization algorithm is stable in the sense that the same clustering is obtained upon repetition. A BIC table, such as the one depicted in Fig. 6.4 is therefore easily obtained.

---

[3]To be more precise, model-based hierarchical clustering (Fraley 1998).

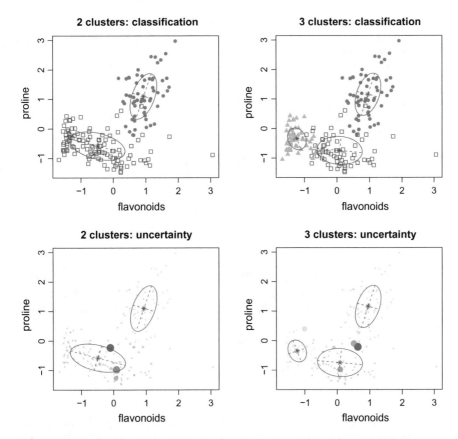

Fig. 6.5 Two- and three-cluster models for the `wines` data, obtained by **mclust**. The top row shows the classifications; the bottom row shows uncertainties at three levels, where the smallest dots have z-values over 0.95 and the largest, black, dots have z-values below 0.75. The others are in between

While mixtures of gaussians (or other distributions) have many attractive properties, they suffer from one big disadvantage: the number of parameters to estimate quickly becomes large. This is the reason why the BIC curve in Fig. 6.4 does not run all the way to nine clusters, although that is the default in `mclust`: in high dimensions, clusters with only few members quickly lead to singular covariance matrices. In such cases, no BIC value is returned. Banfield and Raftery (Banfield and Raftery 1993) suggested to impose restrictions on the covariance matrices of the clusters: one can, e.g., use spherical and equal-sized covariance matrices for all clusters. In this case, which is also the most restricted, the criterion that is optimized corresponds to the criterion used in k-means and in Ward's hierarchical clustering. For each cluster, Gp parameters need to be estimated for the cluster centers, one parameter for the covariance matrices, and p mixing proportions, a total of $(G + 1)p + 1$. In contrast, for the completely free model such as the ones in Figs. 6.4 and 6.5, indicated with

Fig. 6.6 BIC plots for all
covariance models
implemented in `mclust`:
although the constrained
models do not fit as well for
the same numbers of clusters,
they are penalized less and
achieve higher BIC values
for larger numbers of clusters

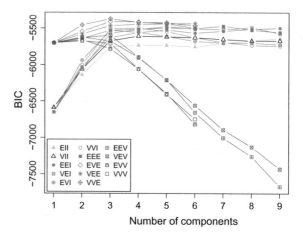

"VVV" in `mclust`, every single covariance matrix requires $p(p + 1)/2$ parameters. This leads to a grand total of $p(Gp + G + 4)/2$ estimates. For low-dimensional data, this is still doable, but for higher-dimensional data the unrestricted models are no longer workable.

Consider the wine data again, but now consider all ten models implemented in `mclust`:

```
> wines.BIC <- mclustBIC(wines.sc)
> plot(wines.BIC, legendArgs = list(x = "bottom", ncol = 2))
```

This leads to the output in Fig. 6.6. The three-letter codes in the legend stand for volume, shape and orientation, respectively. The "E" indicates equality for all clusters, the "V" indicates variability, and the "I" indicates identity. Thus, the "EEI" model stands for diagonal covariance matrices (the "I") with equal volumes and shapes, and the "VEV" model indicates an ellipsoidal model with equal shapes for all clusters, but complete freedom in size and orientation. It is clear that the more constrained models achieve much higher BIC values for higher numbers of clusters: the unconstrained models are penalized more heavily for estimating so many parameters.

6.4 Comparing Clusterings

In many cases, one is interested in comparing the results of different clusterings. This may be to assess the behavior of different methods on the same data set, but also to find out how variable the clusterings are that are obtained by randomly initialized methods like k-means. The difficulty here, of course, is that there is no golden standard; one cannot simply count the number of incorrect assignments and use that as a quality criterion. Moreover, the number of clusters may differ—still we may be interested in assessing the agreement between the partitions.

Several measures have been proposed in literature. Hubert (1985) compares several of these, and proposes the adjusted Rand index, inspired by earlier work by Rand (1971). The original Rand index is based on the number of times two objects are classified in the same cluster, n. In the formulas below, $n_{i\cdot}$ indicates the number of object pairs classified in the same cluster in partition one, but not in partition two, $n_{\cdot j}$ the reverse, and n_{ij} the number of pairs classified in different clusters in both partitions. The index, comparing two partitions with I and J objects, respectively, is given by

$$R = \binom{n}{2} + 2 \sum_{i=1}^{I} \sum_{j=1}^{J} \binom{n_{ij}}{2} - \left\{ \sum_{i=1}^{I} \binom{n_{i\cdot}}{2} + \sum_{j=1}^{J} \binom{n_{\cdot j}}{2} \right\} \qquad (6.3)$$

The adjusted Rand index "corrects for chance" by taking into account the expected value of the index under the null hypothesis of random partitions:

$$R_{\text{adj}} = \frac{R - E(R)}{\max(R) - E(R)} = \frac{a \binom{n}{2} - bc}{\frac{1}{2} \binom{n}{2}(b + c) - bc} \qquad (6.4)$$

with

$$a = \sum_{i,j} \binom{n_{ij}}{2} \qquad (6.5)$$

$$b = \sum_{i} \binom{n_{i\cdot}}{2} \qquad (6.6)$$

$$c = \sum_{j} \binom{n_{\cdot j}}{2} \qquad (6.7)$$

This measure is zero when the Rand index takes its expected value, and has a maximum of one.

The implementation in R takes only a few lines:

```
> AdjRkl <- function(part1, part2) {
+    confusion <- table(part1, part2)
+
+    n <- sum(confusion)
+    a <- sum(choose(confusion[confusion>1], 2))
+    b <- apply(confusion, 1, sum)
+    b <- sum(choose(b[b>1], 2))
+    c <- apply(confusion, 2, sum)
+    c <- sum(choose(c[c>1], 2))
+
+    Rexp <- b*c/choose(n, 2)
+    (a - Rexp) / (.5*(b+c) - Rexp )
+ }
```

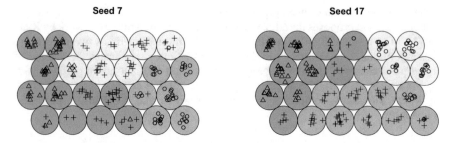

Fig. 6.7 Clustering of the codebook vectors of two mappings of the wine data, indicated by background colors. Symbols indicate vintages

The function takes two partitionings, i.e., class vectors, and returns the value of the adjusted Rand index. Note that the number of classes in both partitionings need not be the same. An alternative is function `adjustedRandIndex` in package **mclust**.

How this can be useful is easily illustrated. As already stated, repeated application of SOM mapping will, in general, lead to mappings that visually can appear very different. However, objects may find themselves very close to the same neighbors in repeated training runs, so that conclusions from the two maps will be very much the same. One way to investigate that is to quantify the similarities. Consider the SOM mapping of the wine data for two initializations:

```
> set.seed(7)
> som.wines <- som(wines.sc, grid = somgrid(6, 4, "hexagonal"))
> set.seed(17)
> som.wines2 <- som(wines.sc, grid = somgrid(6, 4, "hexagonal"))
```

Assessing the similarities of the maps should not be done on the level of the individual units, since these are not relevant entities in themselves. Rather, the units should be aggregated into larger clusters. This can be achieved by looking at plots like Fig. 5.5; an alternative is to explicitly cluster the codebook vectors (see, e.g., Vesanto and Alhoniemi 2000). If hierarchical clustering is used, the dendrograms can be cut at the desired level, immediately providing cluster memberships for the individual samples.

```
> som.hc <- cutree(hclust(dist(getCodes(som.wines, 1))), k = 3)
> som.hc2 <- cutree(hclust(dist(getCodes(som.wines2, 1))), k = 3)
> plot(som.wines, "mapping", bgcol = terrain.colors(3)[som.hc],
+      pch = as.integer(vintages), main = "Seed 7")
> plot(som.wines2, "mapping", bgcol = terrain.colors(3)[som.hc2],
+      pch = as.integer(vintages), main = "Seed 17")
```

This leads to the plots in Fig. 6.7.

The mappings seem very different. Is this really the case, or is it just a visual artifact? Let's find out:

```
> som.clust <- som.hc[som.wines$unit.classif]
> som.clust2 <- som.hc2[som.wines2$unit.classif]
> AdjRkl(som.clust, som.clust2)
[1] 0.4501
```

This rather low value suggests that both mappings are quite different. Note that this analysis does not take into account the vintages and is applicable also in cases where "true" class labels are unknown. Of course, one can also use the adjusted Rand index to compare clusterings with a set of "true" labels:

```
> AdjRkl(vintages, som.clust)
[1] 0.47699
> AdjRkl(vintages, som.clust2)
[1] 0.67278
```

Clearly, the second random seed gives a mapping that is more in agreement with the class labels, something that is also clear when looking at the agreement between plotting symbols and background color in Fig. 6.7.

Other indices to measure correspondence between two partitionings include Fowlkes' and Mallows's B_k (Fowlkes and Mallows 1983), Goodmans and Kruskals γ (Goodman and Kruskal 1954), and Meila's Variation of Information criterion (Meila 2007), also available in **mclust**. The latter is a difference measure, rather than a similarity measure.

6.5 Discussion

Hierarchical clustering methods have many attractive features. They are suitable in cases where there is a hierarchical structure, i.e., subclusters, which very often is the case. A large number of variables does not pose a problem: the rate-limiting step is the calculation of the distance matrix, the size of which does not depend on the dimensionality of the data, but only on the number of samples. And last but not least, the dendrogram provides an appealing presentation of the cluster structure, which can be used to assess clusterings with different numbers of clusters very quickly. Partitional methods, on the other hand, are more general. In hierarchical clustering a split cannot be undone – once a sample is in one branch of the tree, there is no way it can move to the other branch. This can lead, in some cases, to suboptimal clusterings. Partitional methods do not know such restrictions: a sample can always be classified into a different class in the next iteration. Some of the less complicated partitional methods, such as k-means clustering, can also be applied with huge data sets, containing tens of thousands of samples, that cannot be tackled with hierarchical clustering.

Both types of clustering have their share of difficulties, too. In cases relying on distance calculations (all hierarchical methods, and some of the partitional methods, too), the choice of a distance function can dramatically influence the result. The importance of this cannot be overstated. On the one hand, this is good, since it allows one to choose the most relevant distance function available—it even allows one to tackle data that do not consist of real numbers but are binary or have a more complex nature. As long as there is a distance function that adequately represents dissimilarities between objects, the regular clustering methods can be applied. On the other hand, it is bad: it opens up the possibility of a wrong choice. Furthermore, one should realize that when correlated groups of variables are present, as often is the case in life science data, these variables may receive a disproportionly large weight in a regular distance measure such as Euclidean distance, and smaller groups of variables, or uncorrelated variables, may fail to be recognized as important.

Partitional methods force one to decide on the number of clusters beforehand, or perform multiple clusterings with different numbers of clusters. Moreover, there can be considerable differences upon repeated clustering, something that is less prominent in hierarchical clustering (only with ties in the distance data). The main problem with hierarchical clustering is that the bottom-up joining procedure may be too strict: once an object is placed in a certain category, it will stay there, whatever happens further on in the algorithm. Of course, there are many examples where this leads to a sub-optimal clustering. More generally, there may not be a hierarchical structure to begin with.

Both partitional and hierarchical clustering yield "crisp" clusters, that is, objects are assigned to exactly one cluster, without any doubt. For partitional methods, there are alternatives where each object gets a membership value for each of the clusters. If a crisp clustering is required, at the end of the algorithm the object is assigned to the cluster for which it has the highest membership. We have seen one example in the model-based clustering methods.

Finally, one should take care not to over-interpret the results. If you ask for five clusters, that is exactly what you get. Suppose one has a banana-shaped cluster. Methods like k-means, but also complete linkage, will typically describe such a banana with three or four spherical clusters. The question is: are you interested in the peas or the pod[4]? It may very well be that several clusters in fact describe one and the same group, and that to find the other clusters one should actually instruct the clustering to look for more than five clusters.

Clustering is, because of the lack of "hard" criteria, more of an art than a science. Without additional knowledge about the data or the problem, it is hard to decide which one of several different clusterings is best. This, unfortunately, in some areas has led to a practice in which all available clustering routines are applied, and the one that seems most "logical" is selected and considered to describe "reality". One should always keep in mind that this may be a gross overestimation of the powers of clustering.

---

[4]Metaphor from Adrian Raftery.

Part III
Modelling

Chapter 7
Classification

The goal of classification, also known as supervised pattern recognition, is to provide a model that yields the optimal discrimination between several classes in terms of predictive performance. It is closely related to clustering. The difference is that in classification it is clear what to look for: the number of classes is known, and the classes themselves are well-defined, usually by means of a set of examples, the training set. Labels of objects in the training set are generally taken to be error-free, and are typically obtained from information other than the data we are going to use in the model. For instance, one may have data—say, concentration levels of several hundreds of proteins in blood—from two groups of people, healthy, and not-so-healthy, and the aim is to obtain a classification model that distinguishes between the two states on the basis of the protein levels. The diagnosis may have been based on symptoms, medical tests, family history and subjective reasoning of the doctor treating the patient. It may not be possible to distinguish patients from healthy controls on the basis of protein levels, but if one would be able to, it would lead to a simple and objective test.

Apart from having good predictive abilities, an ideal classification method also provides insight in what distinguishes different classes from each other—which variable is associated with an observed effect? Is the association positive or negative? Especially in the natural sciences, this has become an important objective: a gene, protein or metabolite, characteristic for one or several classes, is often called a *biomarker*. Such a biomarker, or more often, set of biomarkers, can be used as an easy and reliable diagnostic tool, but also can provide insight or even opportunities for intervention in the underlying biological processes. Unfortunately, biomarker identification can be extremely difficult. First of all, in cases where the number of variables exceeds the number of cases, it is quite likely that several (combinations of) variables show high correlations with class labels even though there may not be causal relationships. Furthermore, there is a trend towards more complex non-linear modelling methods (often indicated with terms like Machine Learning or Artificial Intelligence) where

© Springer-Verlag GmbH Germany, part of Springer Nature 2020
R. Wehrens, *Chemometrics with R*, Use R!,
https://doi.org/10.1007/978-3-662-62027-4_7

the relationship between variables and outcome can no longer be summarized in simple coefficient values. Hence, interpretation of such models is often impossible.

What is needed as well is a reliable estimate of the success rate of the classifier. In particular, one would like to know how the classifier will perform in the future, on new samples, of course comparable to the ones used in setting up the model. This error estimate is obtained in a validation step—Chap. 9 provides an overview of several different methods. These are all the more important when the classifier of interest has tunable parameters. These parameters are usually optimized on the basis of estimated prediction errors, but as a result the error estimates are positively biased, and a second validation layer is needed to obtain an unbiased error estimate. In this chapter, we will take a simple approach and divide the data in a representative part that is used for building the model (the training set), and an independent part used for testing (the test set). The phrase "independent" is of utmost importance: if, e.g., autoscaling is applied, one should use the column means and standard deviations of the training set to scale the test set. First scaling the complete data set and then dividing the data in training and test sets is, in a way, cheating: one has used information from the test data in the scaling. This usually leads to underestimates of prediction error.

That the training data should be representative seems almost trivial, but in some cases this is hard to achieve. Usually, a random division works well, but also other divisions may be used. In Chap. 4 we have seen that the odd rows of the `wine` data set are very similar to the even rows. In a classification context, we can therefore use the even rows as a training set and the odd rows as a test set:

```
> wines.odd <- seq(1, nrow(wines), by = 2)
> wines.even <- seq(2, nrow(wines), by = 2)
> wines.trn <- wines[wines.odd, ]
> wines.tst <- wines[wines.even, ]
> vint.trn <- vintages[wines.odd]
> vint.tst <- vintages[wines.even]
```

Note that classes are represented proportional to their frequency in the original data in both the training set and the test set. In a couple of cases we will illustrate methods in two dimensions only, looking at the `flavonoids` and `proline` variables in the wine data set:

```
> wines2.trn <- wines.trn[, c(7, 13)]
> wines2.tst <- wines.tst[, c(7, 13)]
```

There are many different ways of using the training data to predict class labels for future data. Discriminant analysis methods use a parametric description of means and covariances. Essentially, observations are assigned to the class having the highest probability density. Nearest-neighbor methods, on the other hand, focus on similarities with individual objects and assign objects to the class that is prevalent in the neighborhood; another way to look at it is to see nearest-neighbor methods as local density estimators. Similarities between objects can also be used directly, e.g., in kernel methods; the most well-known representative of this type of methods is Support Vector Machines (SVMs). A completely different category of classifiers is formed by tree-based approaches. These create a model consisting of a series of binary decisions. Finally, neural-network based classification will be discussed.

Here we will concentrate on the main concepts and show how they should be implemented using standard approaches. Often these are directly supported by modelling methods themselves. An alternative is to use the **caret** package (short for Classification and Regression Training, Kuhn 2008; Kuhn and Johnson 2013), which provides tools for data splitting and validation in the contexts of classification and regression, but also many other topics mentioned in this book. The manual pages and the vignette of the **caret** package provide more information.

7.1 Discriminant Analysis

In discriminant analysis, one assumes normal distributions for the individual classes: $N_p(\mu_k, \Sigma_k)$, where the subscript p indicates that the data are p-dimensional (McLachlan 2004). One can then classify a new object, which can be seen as a point in p-dimensional space, to the class that has the highest probability density ("likelihood") at that point—this type of discriminant analysis is therefore indicated with the term "Maximum-Likelihood" (ML) discriminant analysis.

Consider the following univariate example with two groups (Mardia et al. 1979): group one is $N(0, 5)$ and group 2 is $N(1, 1)$. The likelihoods of classes i are given by

$$L_i (x; \mu_i, \sigma_i) = \frac{1}{\sigma_i \sqrt{2\pi}} \exp\left[-\frac{(x - \mu_i)^2}{2\sigma_i^2} \right] \tag{7.1}$$

It is not too difficult to show that $L_1 > L_2$ if

$$\frac{12}{25}x^2 - x + 1/2 - \ln 5 > 0$$

which in this case corresponds to the regions outside the interval $[-0.9, 2.9]$. In more general terms, one can show (Mardia et al. 1979) that for one-dimensional data $L_1 > L_2$ when

$$x^2 \left(\frac{1}{\sigma_1^2} - \frac{1}{\sigma_2^2} \right) - 2x \left(\frac{\mu_1}{\sigma_1^2} - \frac{\mu_2}{\sigma_2^2} \right) + \left(\frac{\mu_1^2}{\sigma_1^2} - \frac{\mu_2^2}{\sigma_2^2} \right) < 2 \ln \frac{\sigma_2}{\sigma_1} \tag{7.2}$$

This unrestricted form, where every class is individually described with a mean vector and covariance matrix, leads to quadratic class boundaries, and is called "Quadratic Discriminant Analysis" (QDA). Obviously, when $\sigma_1 = \sigma_2$ the quadratic term disappears, and we are left with a linear class boundary—"Linear Discriminant Analysis" (LDA). Both techniques will be treated in more detail below.

Another way of describing the same classification rules is to make use of the Mahalanobis distance:

$$d(\boldsymbol{x}, i) = (\boldsymbol{x} - \boldsymbol{\mu}_i)^T \boldsymbol{\Sigma}_i^{-1} (\boldsymbol{x} - \boldsymbol{\mu}_i) \tag{7.3}$$

Loosely speaking, this expresses the distance of an object to a class center in terms of the standard deviation in that particular direction. Thus, a sample $\boldsymbol{x}$ is simply assigned to the closest class, using the Mahalanobis metric $d(\boldsymbol{x}, i)$. In LDA, all classes are assumed to have the same covariance matrix $\boldsymbol{\Sigma}$, whereas in QDA every class is represented by its own covariance matrix $\boldsymbol{\Sigma}_i$.

7.1.1 Linear Discriminant Analysis

It is easy to show that Eq. 7.2 in the case of two groups with equal variances reduces to

$$|\boldsymbol{x} - \boldsymbol{\mu}_2| > |\boldsymbol{x} - \boldsymbol{\mu}_1| \tag{7.4}$$

Each observation $\boldsymbol{x}$ will be assigned to class 1 when it is closer to the mean of class 1 than of class 2, something that makes sense intuitively as well. Another way to write this is

$$\boldsymbol{\alpha}^T (\boldsymbol{x} - \boldsymbol{\mu}) > 0 \tag{7.5}$$

with

$$\boldsymbol{\alpha} = \boldsymbol{\Sigma}^{-1}(\boldsymbol{\mu}_1 - \boldsymbol{\mu}_2) \tag{7.6}$$

$$\boldsymbol{\mu} = (\boldsymbol{\mu}_1 + \boldsymbol{\mu}_2)/2 \tag{7.7}$$

This formulation clearly shows the linearity of the class boundaries. The separating hyperplane passes through the midpoint between the cluster centers, but is not necessarily perpendicular to the segment connecting the two centers.

In reality, of course, one does not know the true means $\boldsymbol{\mu}_i$ and the true covariance matrix $\boldsymbol{\Sigma}$. One then uses the plugin estimate S, the estimated covariance matrix.[1] In LDA, it is obtained by pooling the individual covariance matrices S_i:

$$S = \frac{1}{n - G} \sum_{i=1}^{G} n_i S_i \tag{7.8}$$

where there are G groups, n_i is the number of objects in group i, and the total number of objects is n.

---

[1] In statistics this is known as the *sample covariance matrix*.

For the wine data, this can be achieved as follows:

```
> wines.counts <- table(vint.trn)
> ngroups <- length(wines.counts)
> wines.groups <- split(as.data.frame(wines.trn), vint.trn)
> wines.covmats <- lapply(wines.groups, cov)
> wines.wcovmats <- mapply('*', wines.covmats, wines.counts,
+                          SIMPLIFY = FALSE)
> wines.pooledcov <-
+   Reduce("+", wines.wcovmats) / (nrow(wines.trn) - ngroups)
```

This piece of code illustrates a convenient feature of the `lapply` function: when the first argument is a vector, it can be used as an index for a function taking also other arguments—here, a list and a vector. Each of the three covariance matrices is multiplied by a weight corresponding to the number of objects in that class. In the final step, the `Reduce` function adds the three weighted covariance matrices. An alternative is to use a plain and simple loop:

```
> wines.pooledcov2 <- matrix(0, ncol(wines), ncol(wines))
> for (i in 1:3) {
+    wines.pooledcov <- wines.pooledcov2 +
+      cov(wines.groups[[i]]) * nrow(wines.groups[[i]])
+ }
> wines.pooledcov2 <-
+    wines.pooledcov2 / (nrow(wines.trn) - ngroups)


> range(wines.pooledcov2 - wines.pooledcov)
[1] 0 0
```

The number of parameters that must be estimated in LDA is relatively small: the pooled covariance matrix contains $p(p + 1)/2$ numbers, and each cluster center p parameters. For G groups this leads to a total of $Gp + p(p + 1)/2$ estimates—for the wine data, with three groups and thirteen variables, this implies 130 estimates.

The LDA classification itself is now easily performed: first we calculate the Mahalanobis distances (using the `mahalanobis` function) to the three class centers using the pooled covariance matrix, and then we determine which of these three is closest for every sample in the training set:

```
> distances <-
+    sapply(1:ngroups,
+           function(i, samples, means, covs)
+              mahalanobis(samples, colMeans(means[[i]]), covs),
+           wines.trn, wines.groups, wines.pooledcov)
> trn.pred <- apply(distances, 1, which.min)
```

Let's compare our predictions with the vintages of the odd-numbered samples:

```
> table(vint.trn, trn.pred)
            trn.pred
vint.trn      1  2  3
  Barbera    24  0  0
  Barolo      0 29  0
  Grignolino  0  0 36
```

The reproduction of the training data is perfect, much better than we have seen with clustering, which is not surprising since the LDA builds the model (in this case the pooled covariance matrix) using the information from the training set with the explicit aim of discriminating between the classes. However, we should not think that future observations are predicted with equal success. The test data should give an indication of what to expect:

```
> distances <-
+    sapply(1:ngroups,
+            function(i, samples, means, covs)
+              mahalanobis(samples, colMeans(means[[i]]), covs),
+            wines.tst, wines.groups, wines.pooledcov)
> tst.pred <- apply(distances, 1, which.min)
> table(vint.tst, tst.pred)
             tst.pred
vint.tst      1  2  3
  Barbera    24  0  0
  Barolo      0 29  0
  Grignolino  1  0 34
```

One Grignolino sample has been classified in the class of the Barbera samples—a very good result, confirming that the problem is not very difficult. Nevertheless, the difference with the unsupervised clustering approaches is obvious.

Of course, R already contains an lda function (in package **MASS**):

```
> wines.ldamod <- lda(wines.trn, grouping = vint.trn,
+                      prior = rep(1, 3)/3)
> wines.lda.testpred <- predict(wines.ldamod, new = wines.tst)
> table(vint.tst, wines.lda.testpred$class)

vint.tst     Barbera Barolo Grignolino
  Barbera        24      0          0
  Barolo          0     29          0
  Grignolino      1      0         34
```

The prior = rep(1, 3)/3 argument in the lda function is used to indicate that all three classes are equally likely a priori. In many cases it makes sense to incorporate information about prior probabilities. Some classes may be more common than others, for example. This is usually reflected in the class sizes in the training set and therefore is taken into account when calculating the pooled covariance matrix, but it is not explicitly used in the discrimination rule. However, it is relatively simple to do

Fig. 7.1 Projection of the training data from the wine data set in the linear discriminant space. It is easy to see that linear class boundaries can be drawn so that all training objects are classified correctly

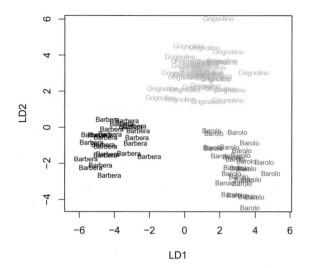

so: instead of maximising L_i one now maximises $\pi_i L_i$, where π_i is an estimate of the prior probability of class i. In the two-group case, this has the effect of shifting the critical value of the discriminant function with an amount of $\log(\pi_2/\pi_1)$ in Eq. 7.5. This approach is sometimes referred to as the Bayesian discriminant rule, and is the default behaviour of the lda function. Obviously, when all prior probabilities are equal, the Bayesian and ML discriminant rules coincide. Also in the example above, using the relative frequencies as prior probabilities would not have made any difference to the predictions—the three vintages have approximately equal class sizes.

The lda function comes with the usual supporting functions for printing and plotting. An example of what the plotting function provides is shown in Fig. 7.1:

```
> plot(wines.ldamod, col = as.integer(vint.trn))
```

The training samples are projected in the space of two new variables, the Linear Discriminants (LDs). In comparison to the PCA scoreplot from Fig. 4.1, class separation has clearly increased. Again, this is the result of the way in which the LDs have been chosen: whereas the PCs in PCA account for as much variance as possible, in LDA the LDs maximize separation. This will be even more clear when we view LDA in the formulation by Fisher (see Sect. 7.1.3).

One particularly attractive feature of the lda function as it is implemented in **MASS** is the possibility to choose different estimators of means and covariances. In particular, the arguments method = "mve" and method = "t" are interesting as they provide robust estimates.

7.1.2 Crossvalidation

When the number of samples is low, there are two important disadvantages of dividing a data set into two parts, one for training and one for testing. The first is that with small test sets, the error estimates are on a very rough scale: if there are ten samples in the test set, the errors are always multiples of ten percent. Secondly, the quality of the model will be lower than it can be: when building the classification model one needs all information one can get, and leaving out a significant portion of the data in general is not helpful. Only with large sets, consisting of, say, tens or hundreds of objects per class, it is possible to create training and test sets in such a way that modelling power will suffer very little while giving a reasonably precise error estimate. Even then, there is another argument against a division into training and test sets: such a division is random, and different divisions will lead to different estimates of prediction error. The differences may not be large, but in some cases they can be important, especially in the case of outliers and/or extremely unlucky divisions.

One solution would be to try a large number of random divisions and to aver-age the resulting estimates. This is indeed a valid strategy—we will come back to this in Chap. 9. A very popular formalization of this principle is called *crossvalida-tion* (Stone 1974). The general procedure is as follows: one leaves out a certain part of the data, trains the classifier on the remainder, and uses the left-out bit—sometimes called the out-of-bag, or OOB, samples—to estimate the error. Next, the two data sets are joined again, and a new test set is split off. This continues until all objects have been left out exactly once. The crossvalidation error in classification is simply the number of misclassified objects divided by the total number of objects in the training set. If the size of the test set equals one, every sample is left out in turn—the procedure has received the name Leave-One-Out (LOO) crossvalidation. It is shown to be unbiased but can have appreciable variance: on average, the estimate is correct, but individual components may deviate considerably.

More stable results are usually obtained by leaving out a larger fraction, e.g., 10% of the data; such a crossvalidation is known as ten-fold crossvalidation. The largest errors cancel out (to some extent) so that the variance decreases; however, one pays the price of a small bias because of the size difference of the real training set and the training set used in the crossvalidation (Efron and Tibshirani 1993). In general, the pros outweigh the cons, so that this procedure is quite often applied. It also leads to significant speed improvements for larger data sets, although for the simple techniques presented in this chapter it is not likely to be very important. The whole crossvalidation procedure is illustrated in Fig. 7.2.

For LDA (and also QDA), it is possible to obtain the LOO crossvalidation result without complete refitting—upon leaving out one object, one can update the Maha-lanobis distances of objects to class means and derive the classifications of the left-out samples quickly, without doing expensive matrix operations (Ripley 1996). The `lda` function returns crossvalidated predictions in the list element `class` when given the argument `CV = TRUE`:

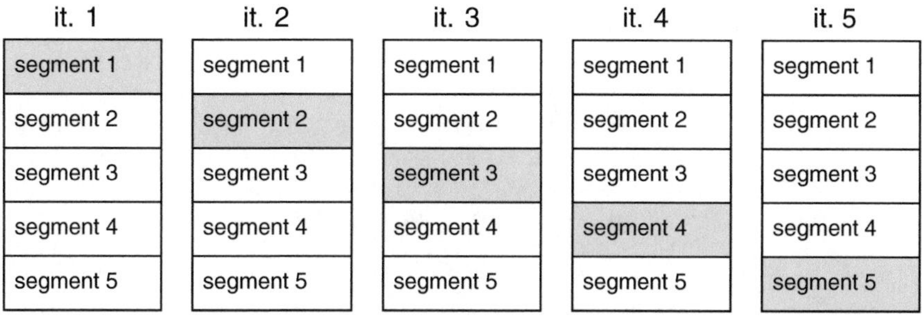

it. 1	it. 2	it. 3	it. 4	it. 5
segment 1	segment 1	segment 1	segment 1	segment 1
segment 2	segment 2	segment 2	segment 2	segment 2
segment 3	segment 3	segment 3	segment 3	segment 3
segment 4	segment 4	segment 4	segment 4	segment 4
segment 5	segment 5	segment 5	segment 5	segment 5

Fig. 7.2 Illustration of crossvalidation; in the first iteration, segment 1 of the data is left out during training and used as a test set. Every segment in turn is left out. From the prediction errors of the left-out samples the overall crossvalidated error estimate is obtained

```
> wines.ldamod <- lda(wines.trn, grouping = vint.trn,
+                      prior = rep(1, 3)/3, CV = TRUE)
> table(vint.trn, wines.ldamod$class)

vint.trn     Barbera Barolo Grignolino
   Barbera        24      0          0
   Barolo          0     28          1
   Grignolino      1      0         35
```

So, where the training set can be predicted without any errors, LOO crossvalidation on the training set leads to an estimated error percentage of $2/89 = 2.25\%$, twice the error on the test set. This difference in itself is not very alarming—error estimates also have variance.

7.1.3 Fisher LDA

A seemingly different approach to discriminant analysis is taken in Fisher LDA, named after its inventor, Sir Ronald Aylmer Fisher. Rather than assuming a particular distribution for individual clusters, Fisher devised a way to find a sensible rule to discriminate between classes by looking for a linear combination of variables a maximizing the ratio of the between-groups sums of squares B and the within-groups sums of squares W (Fisher 1936):

$$a^T Ba/a^T Wa \tag{7.9}$$

These sums of squares are calculated by

$$W = \sum_{i=1}^{G} \widetilde{X}_i^T \widetilde{X}_i \tag{7.10}$$

$$B = \sum_{i=1}^{G} n_i (\bar{x}_i - \bar{x})(\bar{x}_i - \bar{x})^T \tag{7.11}$$

where $\widetilde{X}_i$ is the mean-centered part of the data matrix containing objects of class i, and $\bar{x}_i$ and $\bar{x}$ are the mean vectors for class i and the whole data matrix, respectively. Put differently: W is the variation *around* the class centers, and B is the variation of the class centers around the global mean. It also holds that the total variance T is the sum of the between and within-groups variances:

$$T = B + W \tag{7.12}$$

Fisher's criterion is equivalent to finding a linear combination of variables a corresponding to the subspace in which distances between classes are large and distances within classes are small—compact classes with a large separation. It can be shown that maximizing Eq. 7.9 leads to an eigenvalue problem, and that the solution a is given by the eigenvector of BW^{-1} corresponding with the largest eigenvalue. An object x is then assigned to the closest class, i, which means that for all classes $i \neq j$ the following inequality holds:

$$|a^T x - a^T \bar{x}_i| < |a^T x - a^T \bar{x}_j| \tag{7.13}$$

Interestingly, although Fisher took a completely different starting point and did not explicitly assume normality or equal covariances, in the two-group case Fisher LDA leads to exactly the same solution as ML-LDA. Consider the discrimination between Barbera and Grignolino wines:

```
> wns <- wines[vintages != "Barolo", c(7, 13)]
> vnt <- factor(vintages[vintages != "Barolo"])
> wines.odd2 <- seq(1, nrow(wns), by = 2)
> wines.even2 <- seq(2, nrow(wns), by = 2)
```

To enable easy visualization, we will restrict ourselves to only two variables, flavonoids and proline. Fisher LDA is performed by the following code:

```
> wines.counts <- table(vnt)
> wines.groups <- split(as.data.frame(wns), vnt)
> WSS <-
+    Reduce("+", lapply(wines.groups,
+                       function(x) {
+                          crossprod(scale(x, scale = FALSE))}}))
> BSS <-
+    Reduce("+", lapply(wines.groups,
+                       function(x, y) {
+                          nrow(x) * tcrossprod(colMeans(x) - y)},
+                       colMeans(wns)))
> FLDA <- eigen(solve(WSS, BSS))$vectors[, 1]
> FLDA / FLDA[1]
[1]  1.00000000 -0.00087649
```

Application of ML-LDA, Eq. 7.5, leads to

```
> wines.covmats <- lapply(wines.groups, cov)
> wines.wcovmats <- lapply(1:length(wines.groups),
+                          function(i, x, y) x[[i]]*y[i],
+                          wines.covmats, wines.counts)
> wines.pcov12 <- Reduce("+", wines.wcovmats) / (length(vnt) - 2)
> MLLDA <-
+    solve(wines.pcov12,
+          apply(sapply(wines.groups, colMeans), 1, diff))
> MLLDA / MLLDA[1]
 flavonoids       proline
 1.00000000 -0.00087476
```

Setting the first element of the discrimination functions equal to 1 makes the comparison easier: the vector a in Eq. 7.9 can be rescaled without any effect on both allocation rules Eqs. 7.13 and 7.5. In the two-group case, both ML-LDA and Fisher-LDA lead to the same discrimination function.

For problems with more than two groups, the results are different unless the sample means are collinear: Fisher LDA aims at finding *one* direction discriminating between the classes. An example is shown in Fig. 7.3, where the boundaries between the three classes in the two-dimensional subset of the wine data are shown for Fisher LDA and ML-LDA.

The Fisher LDA boundaries for more than two classes are produced by code essentially identical to the code for the two-group case earlier in this section: one should replace the lines

```
> wns <- wines[vintages != "Barolo", c(7, 13)]
> vnt <- factor(vintages[vintages != "Barolo"])
```

by

```
> wns <- wines2.trn
> vnt <- vint.trn
```

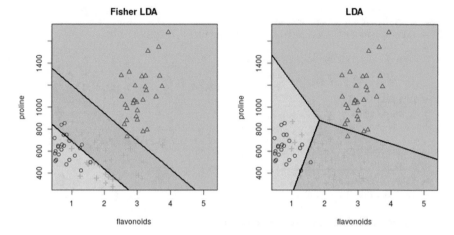

Fig. 7.3 Class boundaries for the wine data (proline and flavonoids only) for Fisher LDA (left) and ML-LDA (right). Models are created using the odd rows of the wine data (training data); the plotting symbols indicate the even rows (test data), as mentioned in the text

Then, after calculating the discriminant function `FLDA`, predictions are made at positions in a regular grid:

```
> xcoo <- seq(.4, 5.4, length = 251)
> ycoo <- seq(250, 1750, length = 251)
> gridXY <- data.matrix(expand.grid(xcoo, ycoo))
> scores <- gridXY %*% FLDA
> meanscores <- c(t(sapply(wines.groups, colMeans)) %*% FLDA)
> Fdistance <- outer(scores, meanscores,
+                    FUN = function(x, y) abs(x - y))
> Fclassif <- apply(Fdistance, 1, which.min)
```

The distances of the scores of all gridpoints to the scores of the class means are calculated using the `outer` function—this leads to a three-column matrix. The classification, corresponding to the class with the smallest distance, is obtained using the function `which.min`.

Finally, the class boundaries are visualized using the functions `image` and `contour`; the points of the test set are added afterwards.

```
> softbrg <- colorRampPalette(c("lightgray", "pink", "lightgreen"))
> image(x = xcoo, y = ycoo,
+       z = matrix(Fclassif, nrow = length(xcoo)),
+       xlab = "flavonoids", ylab = "proline",
+       main = "Fisher LDA", col = softbrg(3))
> box()
```

```
> contour(x = xcoo, y = ycoo, drawlabels = FALSE,
+          z = matrix(Fclassif, nrow = length(xcoo)),
+          add = TRUE)
> points(wines2.tst, col = wine.classes[wines.even],
+          pch = wine.classes[wines.even])
```

The result is shown in the left plot of Fig. 7.3. The right plot is produced analogously:

```
> wines.ldamod <- lda(wines2.trn,
+                          grouping = vint.trn,
+                          prior = rep(1, 3)/3)
> colnames(gridXY) <- colnames(wines)[c(7, 13)]
> lda.2Dclassif <- predict(wines.ldamod, newdata = gridXY)$class
> lda.2DCM <- matrix(as.integer(lda.2Dclassif), nrow = length(xcoo))
> image(x = xcoo, y = ycoo, z = lda.2DCM,
+          xlab = "flavonoids", ylab = "proline",
+          main = "LDA", col = softbrg(3))
> box()
> contour(x = xcoo, y = ycoo, z = lda.2DCM, drawlabels = FALSE,
+          add = TRUE)
> points(wines2.tst, col = wine.classes[wines.even],
+          pch = wine.classes[wines.even])
```

Immediately it is obvious that although the error rates of the two classifications are quite similar for the test set, large differences will occur when data points are further away from the class centers. The class means are reasonably close to a straight line, so that Fisher LDA does not fail completely; however, for multi-class problems it is not a good idea to impose parallel class boundaries, as is done by Fisher LDA using only one eigenvector. It is better to utilize the information in the second and higher eigenvectors of $W^{-1}B$ as well (Mardia et al. 1979); these are sometimes called *canonical variates*, and the corresponding form of discriminant analysis is known as *canonical discriminant analysis*. The maximum number of canonical variates that can be extracted is one less than the number of groups.

7.1.4 Quadratic Discriminant Analysis

Quadratic discriminant analysis (QDA) takes the same route as LDA, with the important distinction that every class is described by its own covariance matrix, rather than one identical (pooled) covariance matrix for all classes. Given our exposé on LDA, the algorithm for QDA is pretty simple: one calculates the Mahalanobis distances of all points to the class centers, and assigns each point to the closest class. Let us see what this looks like in two dimensions:

```
> wines2.groups <- split(as.data.frame(wines2.trn), vint.trn)
> wines2.covmats <- lapply(wines2.groups, cov)
> ngroups <- length(wines2.groups)
> distances <- sapply(1:ngroups,
+                       function(i, samples, means, covs) {
+                          mahalanobis(samples,
+                                        colMeans(means[[i]]),
+                                        covs[[i]]) },
+                       wines2.tst, wines2.groups, wines2.covmats)
> test.pred <- apply(distances, 1, which.min)
> table(vint.tst, test.pred)
            test.pred
vint.tst      1  2  3
  Barbera    19  0  5
  Barolo      0 28  1
  Grignolino  3  1 31
```

Ten samples are misclassified in the test set. To see the class boundaries in two-dimensional space, we use the same visualization as seen in the previous section:

```
> qda.mahal.dists <-
+    sapply(1:ngroups,
+           function(i, samples, means, covs) {
+             mahalanobis(samples,
+                            colMeans(means[[i]]),
+                            covs[[i]]) },
+           gridXY, wines2.groups, wines2.covmats)
> qda.2Dclassif <- apply(qda.mahal.dists, 1, which.min)
> qda.2DCM <- matrix(qda.2Dclassif, nrow = length(xcoo))
> image(x = xcoo, y = ycoo, z = qda.2DCM,
+       xlab = "flavonoids", ylab = "proline",
+       main = "QDA", col = softbrg(3))
> box()
> contour(x = xcoo, y = ycoo, z = qda.2DCM, drawlabels = FALSE,
+          add = TRUE)
> points(wines2.tst, col = as.integer(vint.tst),
+         pch = as.integer(vint.tst))
```

The result is shown in the left plot of Fig. 7.4. The quadratic form of the class boundaries is clearly visible. Again, only the test set objects are shown.

Using the qda function from the **MASS** package, modelling the odd rows and predicting the even rows is done just like with lda. Let's build a model using all thirteen variables:

```
> wines.qda <- qda(wines.trn, vint.trn,
+                   prior = rep(1, 3)/3)
> test.qdapred <- predict(wines.qda, newdata = wines.tst)
> table(vint.tst, test.qdapred$class)
```

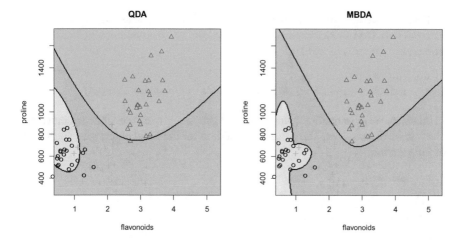

Fig. 7.4 Class boundaries for the wine data (proline and flavonoids only) for QDA (left) and MBDA (right). Models are created using the odd rows of the wine data (training data); the plotting symbols indicate the even rows (test data)

```
vint.tst      Barbera Barolo Grignolino
   Barbera        24      0          0
   Barolo          0     29          0
   Grignolino      0      0         35
```

In this case, all test set predictions are correct.

The optional correction for unequal class sizes (or account for prior probabilities) is done in exactly the same way as in LDA. Several other arguments are shared between the two functions: both lda and qda can be called with the method argument to obtain different estimates of means and variances: the standard plug-in estimators, maximum likelihood estimates, or two forms of robust estimates. The CV argument enables fast LOO crossvalidation.

7.1.5 Model-Based Discriminant Analysis

Although QDA uses more class-specific information, it still is possible that the data are not well described by the individual covariance matrices, e.g., in case of non-normally distributed data. In such a case one can employ more greedy forms of discriminant analysis, utilizing very detailed descriptions of class densities. In particular, one can describe each class with a mixture of normal distributions, just like in model-based clustering, and then assign an object to the class for which the overall mixture density is maximal. Thus, for every class one estimates *several* means and covariance matrices—one describes the pod by a set of peas. Obviously, this technique can only be used when the ratio of objects to variables is very large.

Package **mclust** contains several functions for doing *model-based discriminant analysis* (MBDA), or *mixture discriminant analysis*, as it is sometimes called as well. The easiest one is MclustDA:

```
> wines.MclustDA <-
+    MclustDA(wines.trn, vint.trn, G = 1:5, verbose = FALSE)
```

In this case, we have restricted the number of gaussians, to be used for each individual class, to be at most five. The summary method for the fitted object gives quite a lot of information:

```
> summary(wines.MclustDA)
------------------------------------------------
Gaussian finite mixture model for classification
------------------------------------------------

MclustDA model summary:

 log.likelihood  n  df      BIC
          -1479 89 151 -3635.8

Classes       n Model G
  Barbera    24   VEI 3
  Barolo     29   EEI 3
  Grignolino 36   VEI 2

Training classification summary:

              Predicted
Class      Barbera Barolo Grignolino
  Barbera       24      0          0
  Barolo         0     28          1
  Grignolino     0      0         36

Training error = 0.011236
```

Again, the BIC value is employed to select the optimal model complexity. For the Barolo class, a mixture of three gaussians seems optimal; these all have the same diagonal covariance matrix (indicated with model EEI). The two other classes can be described by mixtures of two and three diagonal covariance matrices, respectively, with varying volume—see Sect. 6.3 for more information on model definition in **mclust**. Classification for the training set is quite successful: only one object is misclassified. Predictions for the test set can be obtained in the usual way:

```
> wines.McDApred <-
+    predict(wines.MclustDA, newdata = wines.tst)$classification
> sum(wines.McDApred != vint.tst)
[1] 1
```

Also here, only one sample is misclassified (a Barolo is seen as a Grignolino).

If more control is needed over the training process, functions `MclustDAtrain` and `MclustDAtest` are available in the **mclust** package. To visualize the class boundaries in the two dimensions of the wine data set employed earlier for the other forms of discriminant analysis, we can use

```
> wines.mclust2D <-
+    MclustDA(wines2.trn, vint.trn, G = 1:5, verbose = FALSE)
```

The model is simpler than the model employed for the full, 13-dimensional data, which seems logical. Prediction and visualization is done by

```
> wines.mclust2Dpred <- predict(wines.mclust2D, gridXY)
> mbda.2DCM <- matrix(as.integer(wines.mclust2Dpred$classification),
+                     nrow = length(xcoo))


> image(x = xcoo, y = ycoo, z = mbda.2DCM,
+        main = "MBDA", xlab = "flavonoids", ylab = "proline",
+        col = softbrg(3))
> box()
> contour(x = xcoo, y = ycoo, z = mbda.2DCM, drawlabels = FALSE,
+         add = TRUE)
> points(wines2.tst,
+        col = as.integer(vint.tst),
+        pch = as.integer(vint.tst))
```

The class boundaries, shown in the right plot of Fig. 7.4, are clearly much more adapted to the densities of the individual classes, compared to the other forms of discriminant analysis we have seen.

7.1.6 Regularized Forms of Discriminant Analysis

At the other end of the scale we find methods that are suitable in cases where we cannot afford to use very complicated descriptions of class density. One form of regularized DA (RDA) strikes a balance between linear and quadratic forms (Friedman 1989): the idea is to apply QDA using covariance matrices $\widetilde{\Sigma}_k$ that are shrunk towards the pooled covariance matrix Σ:

$$\widetilde{\Sigma}_k = \alpha \widehat{\Sigma}_k + (1 - \alpha)\Sigma \tag{7.14}$$

where $\widehat{\Sigma}_k$ is the empirical covariance matrix of class k. In this way, characteristics of the individual classes are taken into account, but they are stabilized by the pooled variance estimate. The parameter α needs to be optimized, e.g., by using crossvalidation.

In cases where the number of variables exceeds the number of samples, more extreme regularization is necessary. One way to achieve this is shrinkage towards the unity matrix (Hastie et al. 2001):

$$\widetilde{\Sigma} = \alpha\Sigma + (1 - \alpha)I \tag{7.15}$$

Equivalent formulations are given by:

$$\widetilde{\Sigma} = \kappa\Sigma + I \tag{7.16}$$

and

$$\widetilde{\Sigma} = \Sigma + \kappa I \tag{7.17}$$

with $\kappa \geq 0$. In this form of RDA, again the regularized form $\widetilde{\Sigma}$ of the covariance is used, rather than the empirical pooled estimate Σ. Matrix $\widetilde{\Sigma}$ is not singular so that the matrix inversions in Eqs. 7.3 or 7.6 no longer present a problem. In the extreme case, one can use a diagonal covariance matrix (with the individual variances on the diagonal) leading to diagonal LDA (Dudoit et al. 2002), also known as Idiot's Bayes (Hand and Yu 2001). Effectively, all dependencies between variables are completely ignored. For so-called "fat" matrices, containing many more variables than objects, often encountered in microarray research and other fields in the life sciences, such simple methods often give surprisingly good results.

7.1.6.1   Diagonal Discriminant Analysis

As an example, consider the odd rows of the prostate data, limited to the first 1000 variables. We are concentrating on the separation between the control samples and the cancer samples:

```
> prostate <- rowsum(t(Prostate2000Raw$intensity),
+                    group = rep(1:327, each = 2),
+                    reorder = FALSE) / 2
> prostate.type <- Prostate2000Raw$type[seq(1, 654, by = 2)]
>
> prost <- prostate[prostate.type != "bph", 1:1000]
> prost.type <- factor(prostate.type[prostate.type != "bph"])
> prost.df <- data.frame(type = prost.type, prost = prost)
> prost.odd <- seq(1, length(prost.type), by = 2)
> prost.even <- seq(2, length(prost.type), by = 2)
```

Although it is easy to re-use the code given in Sects. 7.1.1 and 7.1.4, plugging in diagonal covariance matrices, here we will use the dDA function from the **sfsmisc** package:

```
> prost.dlda <-
+   dDA(prost[prost.odd, ], as.integer(prost.type)[prost.odd])
```

By default, the same covariance matrix is used for all classes, just like in LDA. Here, the result for the predictions on the even samples is not too bad:

```
> prost.dldapred <- predict(prost.dlda, prost[prost.even, ])
> table(prost.type[prost.even], prost.dldapred)
          prost.dldapred
            1  2
  control 32  8
  pca      7 77
```

Approximately 88% of the test samples are predicted correctly. Allowing for different covariance matrices per class, we arrive at diagonal QDA, which does slightly worse for these data:

```
> prost.dqda <-
+    dDA(prost[prost.odd, ], as.integer(prost.type)[prost.odd],
+        pool = FALSE)
> prost.dqdapred <- predict(prost.dqda, prost[prost.even, ])
> table(prost.type[prost.even], prost.dqdapred)
          prost.dqdapred
            1  2
  control 38  2
  pca     16 68
```

7.1.6.2 Shrunken Centroid Discriminant Analysis

In the context of microarray analysis, it has been suggested to combine RDA with the concept of "shrunken centroids" (Tibshirani et al. 2003)—the resulting method is indicated as SCRDA (Guo et al. 2007) and is available in the R package **rda**. As the name suggests, class means are shrunk towards the overall mean. The effect is that the points defining the class boundaries (the centers) are closer, which may lead to a better description of local structure. These shrunken class means are then used in Eq. 7.3, together with the diagonal covariance matrix also employed in DLDA. For a more complete description, see, e.g., (Hastie et al. 2001).

Let us see how SCRDA does on the prostate example. Application of the rda function is straightforward.[2] The function takes two arguments, α and δ, where α again indicates the amount of unity matrix in the covariance estimate, and δ is a soft threshold, indicating the minimal coefficient size for variables to be taken into account in the classification:

```
> prost.rda <-
+    rda(t(prost[prost.odd, ]), as.integer(prost.type)[prost.odd],
+        delta = seq(0, .4, len = 5), alpha = seq(0, .4, len = 5))
```

Printing the fitted object shows some interesting results:

---

[2]Note that in this function the variables are in the *rows* of the data matrix and not, as usual, in the columns—hence the use of the transpose function.

```
> prost.rda
Call:
rda(x = t(prost[prost.odd, ]), y = as.integer(prost.type)[prost.odd],
    alpha = seq(0, 0.4, len = 5), delta = seq(0, 0.4, len = 5))
$nonzero
       delta
alpha     0 0.1 0.2 0.3 0.4
  0    1000 433 193 121  92
  0.1 1000 220  34   3   0
  0.2 1000 192  19   4   0
  0.3 1000 179  18   4   0
  0.4 1000 195  24   4   0

$errors
       delta
alpha  0 0.1 0.2 0.3 0.4
  0   36  38  39  39  39
  0.1 10  16  32  41  41
  0.2  7  21  43  41  41
  0.3  4  23  44  41  41
  0.4  2  20  44  41  41
```

Increasing values of δ lead to a rapid decrease in the number of non-zero coefficients; however, these sparse models do not lead to very good predictions, and the lowest value for the training error is found at $\alpha = .4$ and $\delta = 0$. Obviously, the training error is not the right criterion to decide on the optimal values for these parameters. This we can do using the rda.cv crossvalidation function, and subsequently we can use the test data as a means to estimate the expected prediction error:

```
> prost.rdacv <-
+    rda.cv(prost.rda, t(prost[prost.odd, ]),
+           as.integer(prost.type)[prost.odd])
```

Inspection of the result (not shown) reveals that the optimal value of α would be .2, with no thresholding (*delta* $= 0$). Predictions with these values lead to the following result:

```
> prost.rdapred <-
+    predict(prost.rda,
+            t(prost[prost.odd, ]), as.integer(prost.type)[prost.odd],
+            t(prost[prost.even, ]), alpha = .2, delta = 0)
> table(prost.type[prost.even], prost.rdapred)
          prost.rdapred
            1  2
  control  30 10
  pca       4 80
```

Overall, fourteen samples are misclassified, only slightly better than the DLDA model. This sort of behavior is more general than one might think: for fat data, the simplest models are often among the top performers.

7.2 Nearest-Neighbor Approaches

A completely different approach, not relying on any distributional assumptions what-soever, is formed by techniques focusing on distances between objects, and in partic-ular on the closest objects. These techniques are known under the name of k-nearest-neighbors (KNN), where k is a number to be determined. If $k = 1$, only the closest neighbor is taken into account, and any new object will be assigned to the class of its closest neighbor in the training set. If $k > 1$, the classification is straightforward in cases where the k nearest neighbors are all of the same class. If not, a majority vote is usually performed. Class areas can be much more fragmented than with LDA or QDA; in extreme cases one can even find patch-work-like patterns. The smaller the number k, the more irregular the areas can become: only one object is needed to assign its immediate surroundings to a particular class.

As an example, consider the KNN classification for the first sample in the test set of the wine data (sample number two), based on the training set given by the odd samples. One starts by calculating the distance to all samples in the training set. Usually, the Euclidean distance is used—in that case, one should scale the data appropriately to avoid large numbers to dominate the results. For the wine data, autoscaling is advisable. The `mahalanobis` function has a useful feature that allows one to calculate the distance of one object to a set of others. The covariance matrix is given as the third argument. Thus, the Euclidean distance between samples in the autoscaled wine data can be calculated in two ways, either from the autoscaled data using a unit covariance matrix, or from the unscaled data using the estimated column standard deviations. Consider the wine data, scaled according to the means and variances of the odd rows (the training set). Calculating the distance to the second sample in these two ways leads to the following result:

```
> wines.trn.sc <- scale(wines.trn)
> wines.tst.sc <- scale(wines.tst,
+                        scale = apply(wines.trn, 2, sd),
+                        center = colMeans(wines.trn))


> dist2sample2a <- mahalanobis(wines.trn.sc,
+                              wines.tst.sc[1, ],
+                              diag(13))
> dist2sample2b <- mahalanobis(wines.trn,
+                              wines.tst[1, ],
+                              diag(apply(wines.trn, 2, var)))
>
> range(dist2sample2a - dist2sample2b)
[1] -7.1054e-15  1.4211e-14
```

Clearly, the two lead to the same result. Next, we order the training samples according to their distance to the second sample:

```
> nearest.classes <- vint.trn[order(dist2sample2a)]
> table(nearest.classes[1:10])

  Barbera      Barolo Grignolino
        0          10          0
```

The closest ten objects are all of the Barolo class—apparently, there is little doubt that object 2 also should be a Barolo.

Rather than using a diagonal of the covariance matrix, one could also use the complete estimated covariance matrix of the training set. This would lead to the Mahalanobis distance:

```
> dist2sample2 <- mahalanobis(wines.trn,
+                             wines.tst[1, ],
+                             cov(wines.trn))
> nearest.classes <- vint.trn[order(dist2sample2)]
> table(nearest.classes[1:10])

  Barbera      Barolo Grignolino
        0           6          4
```

Note that autoscaling of the data is not necessary because we explicitly include the covariance matrix of the training data. Clearly, the results depend on the distance measure employed. Although the closest three samples are Barolo wines, the next three are Grignolinos; values of k between 5 and 9 would lead to a close call or even a tie. Several different strategies to deal with such cases can be employed. The simplest is to require a significant majority for any classification—in a 5-NN classification one may require at least four of the five closest neighbors to belong to the same class. If this is not the case, the classification category becomes "unknown". Although this may seem a weakness, in many applications it is regarded as a strong point if a method can indicate some kind of reliability—or lack thereof—for individual predictions.

The **class** package contains an implementation of the KNN classifier using Euclidean distances, knn. Its first argument is a matrix constituting the training set, and the second argument is the matrix for which class predictions are required. The class labels of the training set are given in the third argument. It provides great flexibility in handling ties: the default strategy is to choose randomly between the (tied) top candidates, so that repeated application can lead to different results:

```
> knn(wines.sc[wines.odd, ], wines.sc[68, ], cl = vint.trn, k = 4)
[1] Barbera
Levels: Barbera Barolo Grignolino
> knn(wines.sc[wines.odd, ], wines.sc[68, ], cl = vint.trn, k = 4)
[1] Grignolino
Levels: Barbera Barolo Grignolino
```

Apparently, there is some doubt about the classification of sample 68—it can be either a Barbera or Grignolino. Of course, this is caused by the fact that from the four closest neighbors, two are Barberas and two are Grignolinos. Requiring at least three votes for an unambiguous classification ($1 = 3$) leads to:

```
> knn(wines.sc[wines.odd, ], wines.sc[68, ], cl = vint.trn,
+      k = 4, l = 3)
[1] <NA>
Levels: Barbera Barolo Grignolino
```

In many cases it is better not to have a prediction at all, rather than a highly uncertain one.

The value of k is crucial. Unfortunately, no rules of thumb can be given on the optimal choice, and this must be optimized for every data set separately. One simple strategy is to monitor the performance of the test set for several values of K and pick the one that leads to the smallest number of misclassifications. Alternatively, LOO crossvalidation can be employed:

```
> wines.knnresult <- rep(0, 10)
> for (i in 1:10) {
+    wines.knncv <- knn.cv(wines.sc[wines.odd, ], vint.trn, k = i)
+    wines.knnresult[i] <-
+       sum(diag(table(vint.trn, wines.knncv)))
+ }
> round(100 * wines.knnresult / length(wines.odd), 1)
 [1] 92.1 92.1 96.6 92.1 95.5 96.6 95.5 95.5 96.6 94.4
```

In this example, k values of three, six and nine show the best prediction—differences are not large.

An alternative is to use the convenience function `tune.knn` in package **e1071**. This function by default uses ten-fold crossvalidation for a range of values of k:

```
> (knn.tuned <- tune.knn(wines.sc[wines.odd, ], vint.trn, k = 1:10))

Parameter tuning of 'knn.wrapper':

- sampling method: 10-fold cross validation

- best parameters:
 k
 3

- best performance: 0.022222
```

Plotting the `knn.tuned` object leads to the left plot of Fig. 7.5—the differences with the LOO results we saw earlier show what kind of variability is to be expected with crossvalidated error estimates. Indeed, repeated application of the tune function will—for these data—lead to quite different estimates for the optimal value of k:

```
> bestKs <- rep(0, 1000)
> for (i in 1:1000)
+    bestKs[i] <- tune.knn(wines.sc[wines.odd, ],
+                          vint.trn,
+                          k = 1:10)$best.parameters[1, 1]
```

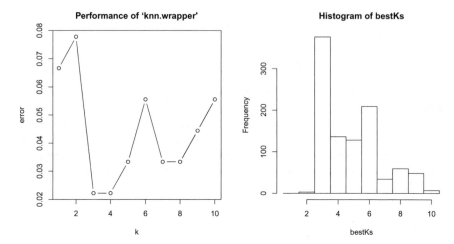

Fig. 7.5 Optimization of k for the wine data using the `tune` wrapper function. Left plot: one crossvalidation curve. Right plot: optimal values of k in 1000 crossvalidations

The right plot of Fig. 7.5 shows a histogram of the best k values. In the large majority of cases, $k = 2$ is best. This is partly caused by a built-in preference for small values of k in the script: the smallest value of k that gives the optimal predictions is stored, even though larger values may lead to equally good predictions, something that given the rather small size of our data set can easily occur.

Although application of these simple strategies allow one to choose the optimal parameter settings, the optimal error associated with this setting (e.g., 97.8% in the LOO example) is not an estimation of the prediction error of future samples, because the test set is used in this procedure to fine-tune the method. Another layer of validation is necessary to find the estimated prediction error; see Chap. 9. The 1-nearest neighbor method enjoys great popularity, despite coming out worst in the above comparison—there, it is almost never selected. Nevertheless, it has been awarded a separate function in the **class** package: `knn1`. Most often, odd values of K smaller than ten are considered.

One potential disadvantage of the KNN method is that in principle, the whole training set—the training set in a sense *is* the model!—should be saved, which can be a nuisance for large data sets. Predictions for new objects can be slow, and storing really large data sets may present memory problems. However, things are not so bad as they seem, since in many cases one can safely prune away objects without sacrificing information. For the wine data, it is obvious that in large parts of the space there is no doubt: only objects from one class are present. Many of these objects can be removed and one then still will get exactly the same classification for all possible new objects.

7.3 Tree-Based Approaches

A wholly different approach to classification is formed by tree-based approaches. These proceed in a way that is very similar to medical diagnosis: the data are "interrogated" and a series of questions are posed which finally lead to a classification. Modelling, in this metaphor, is to decide which questions are most informative. As a class, tree-based methods possess some unique advantages. They can be used for both classification and regression. Since the model is based on sequential decisions on individual variables, scaling is not important: every variable is treated at its own scale and no "overall" measure needs to be computed. Variable selection comes with the method—only those variables that contribute to the result are incorporated in the tree. Trees form one of the few methods that can accommodate variables of a very different nature, e.g., numerical, categorical and ordinal, in one single model. Their handling of missing values is simple and elegant. In short, trees can be used for almost any classification (and regression) problem.

Currently, tree-based modelling comes in two main branches: Breiman's Classification and Regression Trees (CART, Breiman et al. 1984) and Quinlan's See5/C5.0 (and its predecessors, C4.5, Quinlan 1993, and ID3, Quinlan 1986). Both are commercial and not open-source software, but R comes with two pretty faithful representation of CART in the form of the `rpart` and `tree` functions, from the packages with the same names. Since the approaches by Quinlan and Breiman have become more similar with every new release, and since no See5/C5.0 implementation is available in R, we will here only focus on `rpart` (Therneau and Atkinson 1997)—one of the recommended R packages—to describe the main ideas of tree-based classification. The differences between CART and the implementation in the **tree** package are small; consult the manual pages and (Therneau and Atkinson 1997) for more information.

7.3.1 Recursive Partitioning and Regression Trees

Recursive Partitioning and Regression Trees, which is what the acronym `rpart` stands for, can be explained most easily by means of an example. Consider, again, the the two-dimensional subset of the wine data encountered earlier:

```
> wines2.df <- data.frame(vint = vintages, wines[, c(7, 13)])
> wines2.rpart <- rpart(vint ~ ., subset = wines.odd,
+                       data = wines2.df, method = "class")
```

In setting up the tree model, we explicitly indicate that we mean classification (`method = "class"`): the `rpart` function also provides methods for survival analysis and regression, and though it tries to be smart in guessing what exactly is required, it is better to explicitly provide the `method` argument. The result is an object of class `rpart`:

```
> wines2.rpart
n= 89

node), split, n, loss, yval, (yprob)
      * denotes terminal node

1) root 89 53 Grignolino (0.269663 0.325843 0.404494)
  2) flavonoids< 1.235 23  1 Barbera (0.956522 0.000000 0.043478) *
  3) flavonoids>=1.235 66 31 Grignolino (0.030303 0.439394 0.530303)
    6) proline>=739 30  2 Barolo (0.000000 0.933333 0.066667) *
    7) proline< 739 36  3 Grignolino (0.055556 0.027778 0.916667) *
```

The top node, where no splits have been defined, is labelled as "Grignolino" since
that is the most abundant variety—36 out of 89 objects (a fraction of 0.4045) are
Grignolinos. The first split is on the flavonoids variable. A value smaller than
1.235 leads to node 2, which is consisting almost completely of Barbera samples
(more than 95 percent). This node is not split any further, and in tree terminology
is indicated as a "leaf". A flavonoid value larger than 1.235 leads to node three that
is split further into separate Barolo and Grignolino leaves. Of course, such a tree is
much easier to interpret when depicted graphically:

```
> plot(wines2.rpart, margin = .12)
> text(wines2.rpart, use.n = TRUE)
> cl.id <- as.integer(wines2.df$vint[wines.odd])
> plot(wines2.df[wines.odd, 2:3], pch = cl.id, col = cl.id)
> segments(wines2.rpart$splits[1, 4], par("usr")[3],
+           wines2.rpart$splits[1, 4], par("usr")[4], lty = 2)
> segments(wines2.rpart$splits[1, 4], wines2.rpart$splits[2, 4],
+           par("usr")[2], wines2.rpart$splits[2, 4], lty = 2)
```

This leads to the plots in Fig. 7.6. The tree on the left shows the splits, corresponding
to the tessellation of the surface in the right plot. The plot command sets up the
coordinate system en plots the tree; the margin argument is necessary to reserve
some space for the annotation added by the text command. At every split, the test,
stored in the splits element in the X.rpart object, is shown. At the final nodes
(the "leaves"), the results are summarized: one Barbera (red triangles) is classified as
a Grignolino (green pluses), two Barolos (black circles) are in the Grignolino area;
two Grignolinos are thought to be Barolos, one a Barbera.

The rpart function has a familiar formula interface also used in, e.g., lm. For
the wine data, we will predict the classes of the even rows, again based on the odd
rows:

```
> wines.df <- data.frame(vint = vintages, wines)
> wines.rpart <- rpart(vint ~ ., subset = wines.odd,
+                       data = wines.df, method = "class")
```

Plotting the rpart object leads to Fig. 7.7. The flavonoids and proline variables are
again important in the classification, now in addition to the colour intensity.

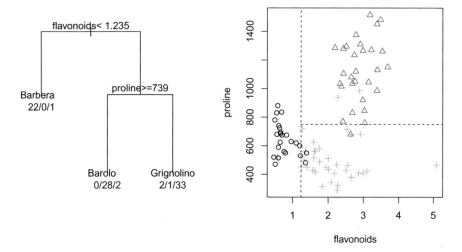

Fig. 7.6 Tree object from `rpart` using default settings (left). The two nodes lead to the class boundaries visualized in the right plot. Only points for the even rows, the test set, are shown

Prediction is done using the `predict.rpart` function, which returns a matrix of class probabilities, simply estimated from the composition of the training samples in the end leaves:

```
> wines.rpart.predict <- predict(wines.rpart,
+                                newdata = wines.df[wines.even, ])
> wines.rpart.predict[31:34, ]
    Barbera    Barolo Grignolino
62        0 0.032258    0.96774
64        0 0.032258    0.96774
66        0 0.142857    0.85714
68        0 0.032258    0.96774
```

In this rather simple problem, most of the probabilities are either 0 or 1, but here some Grignolinos are shown that have a slight chance of actually being Barolos, according to the tree model. The uncertainties are simply the misclassification rates of the training model: row 66 ends up in a lead containing seven training samples, one of which is a Barolo and six are Grignolinos. The other rows end up in the large Grignolino group, containing also one Barolo sample. A global overview is more easily obtained by plotting the probabilities:

```
> matplot(wines.rpart.predict, xlab = "sample number (test set)")
```

This leads to the plot in Fig. 7.8. Clearly, most of the Barolos and Barberas are classified with complete confidence, corresponding with "pure" leaves. The Grignolinos on the other hand always end up in a leaf also containing one Barolo sample. When using the `type = "class"` argument to the prediction function, the result is immediately expressed in terms of classes, and can be used to assess the overall prediction quality:

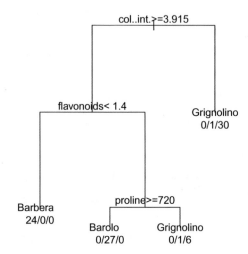

Fig. 7.7 Fitted tree using `rpart` on the odd rows of the `wine` data set (all thirteen variables)

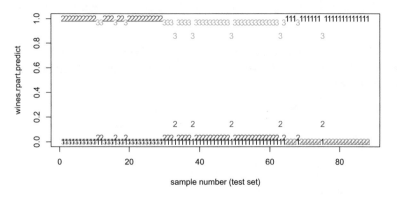

Fig. 7.8 Classification probabilities of the test set of the wine data for the tree shown in Fig. 7.7. Within the first twenty samples we see four incorrect predictions: the true class is Barolo (indicated by "2") but there is some confusion with Grignolino ("3"). Similarly, in the right part of the plot two Barbera samples are seen as Grignolinos

```
> table(vint.tst,
+         predict(wines.rpart, newdata = wines.df[wines.even, ],
+             type = "class"))

vint.tst     Barbera Barolo Grignolino
  Barbera        22      0          2
  Barolo          0     25          4
  Grignolino      0      0         35
```

This corresponds to the six misclassifications seen in Fig. 7.8.

7.3.1.1 Constructing the Tree

The construction of the optimal tree cannot be guaranteed to finish in polynomial time (a so-called NP-complete problem), and therefore one has to resort to simple approximations. The standard approach is the following. All possible splits—binary divisions of the data—in all predictor variables are considered; the one leading to the most "pure" branches is selected. The term "pure" in this case signifies that, in one leaf, only instances of one class are present. For categorical variables, tests for unique values are used; for continuous variables, all data points are considered as potential split values. This simple procedure is applied recursively until some stopping criterion is met.

The crucial point is the definition of "impurity": several different measures can be used. Two criteria are standing out (Ripley 1996): the Gini index, and the entropy. The Gini index of a node is given by

$$I_G(p) = \sum_{i \neq j} p_i p_j = 1 - \sum_j p_j^2 \tag{7.18}$$

and is minimal (exactly zero) when the node contains only samples from one class— p_i is the fraction of samples from class i in the node. A simple function calculating the Gini index looks like this:

```
> gini <- function(x, clss) {
+     p <- table(clss) / length(clss)
+     gini.parent <- 1 - sum(p^2)
+
+     gini.index <-
+         sapply(sort(x), function(splitpoint) {
+             left.ones <- clss[x < splitpoint]
+             right.ones <- clss[x >= splitpoint]
+             nleft <- length(left.ones)
+             nright <- length(right.ones)
+
+             if ((nleft == 0) | (nright == 0)) return (NA)
+
+             p.left <- table(left.ones) / nleft
+             p.right <- table(right.ones) / nright
+
+             (nleft * (1 - sum(p.left^2)) +
+              nright * (1 - sum(p.right^2))) /
+               (nleft + nright)
+         })
+
+     gini.index - gini.parent
+ }
```

This function takes a vector x, for instance values for the proline variable in the wines data, and a class vector. Impurity values are calculated where each value in the sorted vector x is considered as a split point. To really quantify improvement after splitting at that node, the Gini index of the parent node is subtracted: the more negative the number, the bigger the improvement.

The other impurity criterion is based on entropy, where the entropy of a node is defined by

$$I_E(p) = -\sum_j p_j \log p_j$$

which again is minimal when the node is pure and contains only samples of one class (where we define $0 \log 0 = 0$).

The optimal split is the one that minimizes the average impurity of the new left and right branches (whatever criterion is used):

$$P_l I(p_l) + P_r I(p_r)$$

where P_l and P_r signify the sample fractions and $I(p_l)$ and $I(p_r)$ are the impurities of the left and right branches, respectively.

As an illustration, again consider the two-dimensional subset of the odd rows of the wine data, using variables flavonoids and proline. Since the data are continuous, we consider all values as potential splits, and calculate the Gini and entropy indices. For the two-dimensional wine data this leads to:

```
> wines2.df.odd <- wines2.df[wines.odd, ]
> Ginis <- sapply(wines2.df.odd[, -1], gini, wines2.df.odd$vint)
> apply(Ginis, 2, min, na.rm = TRUE)
flavonoids      proline
  -0.24683      -0.24127
> (idx <- which.min(Ginis[, 1]))
[1] 24
> (bestSplit <- sort(wines2.df.odd[, "flavonoids"])[idx])
[1] 1.25
```

Plotting the Gini values for the two columns leads to the left panel in Fig. 7.9. Because of the lower Gini index, corresponding to more pure leaves, the first split should be on the flavonoids column. The split point equals the 24th sorted value. Next we divide the data into two sets, one to the left of the bestSplit value, and one to the right. Here we show only the result for the right split following results:

```
> wr <- wines2.df.odd[wines2.df.odd$flavonoids >= bestSplit, ]
> GinisR <- sapply(wr[, -1], gini, wr$vint)
> apply(GinisR, 2, min, na.rm = TRUE)
flavonoids      proline
  -0.18750      -0.38321
> (idxR <- which.min(GinisR[, 2]))
[1] 37
> (bestSplitR <- sort(wr[, "proline"])[idxR])
[1] 760
```

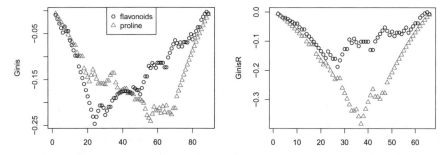

Fig. 7.9 Impurity values (Gini indices) for all possible split points in the two-dimensional subset of the wine data. The left panel points to the `flavonoids` variable for selecting the first split point; the right panel shows that the subsequent fit with the biggest gain is in the `proline` variable

Clearly, the second split should be done for the `proline` column at the level 760. The two splits correspond exactly to the results in Fig. 7.6. The Gini values for the right split are shown in the right panel of Fig. 7.9.

Obviously, one can keep on splitting nodes until every sample in the training set is a leaf in itself, or in any case until each single leaf contains only instances of one class. Such a tree is able to represent the training data perfectly, but whether the predictions of such a tree are reliable is quite another matter. In fact, these trees generally will not perform very well. By describing every single feature of the training set, the tree is not able to generalize. This is an example of *overfitting* (or overtraining, as it is sometimes called as well), something that is likely to occur in methods that have a large flexibility—in the case of trees, the freedom to keep on adding nodes.

The way this problem is tackled in constructing optimal trees is to use *pruning*, i.e., trimming useless branches. When exactly a branch is useless needs to be assessed by some form of validation—in **rpart**, tenfold crossvalidation is used by default. One can therefore easily find out whether a particular branch leads to a decrease in prediction error or not.

More specifically, in pruning one minimizes the cost of a tree, expressed as

$$C(T) = R(T) + \alpha|T| \tag{7.19}$$

In this equation, T is a tree with $|T|$ leafs, $R(T)$ the "risk" of the tree—e.g., the proportion of misclassifications—and α a complexity penalty, chosen between 0 and ∞. One can see α as the cost of adding another node. It is not necessarily to build up the complete tree to calculate this measure: during the construction the cost of the current tree can be assessed and if it is above a certain value, the process stops. This cost is indicated with the complexity parameter (`cp`) in the `rpart` function, which is normalized so that the root node has a complexity value of one.

Once again looking at the first 1000 variables of the `control` and `pca` classes in the prostate data, one can issue the following commands to construct the full tree with no misclassifications in the training set:

```
> prost.df <- data.frame(type = prost.type, prost = prost)
> prost.rpart <-
+    rpart(type ~ ., data = prost.df, subset = prost.odd,
+           control = rpart.control(cp = 0, minsplit = 0))
```

The two extra arguments tell the `rpart` function to keep on looking for splits even when the complexity parameter, `cp`, gets smaller than 0.1 and the minimal number of objects in a potentially split node, `minsplit`, is smaller than 20 (the default values). This leads to a tree with seven leaves. Printing the `prost.rpart` object would show that the training data are indeed predicted perfectly. However, four of the terminal nodes contain only three or fewer samples: it seems these are introduced to repair some individual cases. Indeed, the test data are not predicted with the same level of accuracy:

```
> prost.rpartpred <-
+    predict(prost.rpart, newdata = prost.df[prost.even, ])
> table(prost.type[prost.even], classmat2classvec(prost.rpartpred))

          control pca
  control      29  11
  pca          12  72
```

Pruning could decrease the complexity without sacrificing much accuracy in the description of the training set, and hopefully would increase the generalizing abilities of the model. To see what level of pruning is necessary, the table of complexity values can be printed:

```
> printcp(prost.rpart)

Classification tree:
rpart(formula = type ~ ., data = prost.df, subset = prost.odd,
    control = rpart.control(cp = 0, minsplit = 0))

Variables actually used in tree construction:
[1] prost.4909 prost.5013 prost.5110 prost.5261 prost.5489 prost.5866

Root node error: 41/125 = 0.328

n= 125

        CP nsplit rel error xerror  xstd
1 0.5610      0    1.0000  1.000 0.128
2 0.2927      1    0.4390  0.829 0.121
3 0.0488      2    0.1463  0.634 0.111
4 0.0244      4    0.0488  0.561 0.106
5 0.0000      6    0.0000  0.561 0.106
```

Also a graphical representation is available:

```
> plotcp(prost.rpart)
```

Fig. 7.10 Complexity
pruning of a tree: in this case,
three terminal nodes are
optimal (lowest prediction
error at lowest complexity)

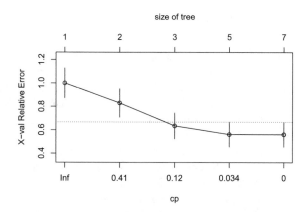

This leads to Fig. 7.10. Both from this figure and the complexity table shown above, it is clear that the tree with the lowest prediction error and the least number of nodes is obtained at a value of cp equal to 0.12. Usually, one chooses the complexity corresponding to the minimum of the predicted error plus one standard deviation, indicated by the dotted line in Fig. 7.10. The tree created with a cp value of 0.12, containing only two leaves rather than the original seven, leads to a higher number of misclassifications (six rather than zero) in the training set, but unfortunately also to a slightly higher number of misclassifications in the test set:

```
> prost.rpart2 <-
+    rpart(type ~ ., data = prost.df, subset = prost.odd,
+         control = rpart.control(cp = 0.12))
> prost.rpart2pred <-
+    predict(prost.rpart2, newdata = prost.df[prost.even, ])
> table(prost.type[prost.even], classmat2classvec(prost.rpart2pred))

         control pca
 control      29  11
 pca          15  69
```

Either way, the result is quite a bit worse than what we have seen earlier with RDA (Sect. 7.1.6.2).

Apart from the 0/1 loss function normally used in classification (a prediction is either right or wrong), **rpart** allows to specify other, more complicated loss functions as well—often, the cost of a false positive is very different from the cost of a false negative decision. Another useful feature in the **rpart** package is the possibility to provide prior probabilities for all classes.

7.3.2 Discussion

Trees offer a lot of advantages. Perhaps the biggest of them is the appeal of the particular form of the model: many scientists feel comfortable with a series of more and more specific questions, eventually leading to an unambiguous answer. The implicit variable selection makes model interpretation much easier, and alleviates many problems with missing values, and variables of mixed types (boolean, categorical, ordinal, numerical).

There are downsides too, of course. The number of parameters to adjust is large, and although the default settings quite often lead to reasonable solutions, there may be a temptation to keep fiddling until an even better result is obtained. This, however, can easily lead to overfitting: although the data are faithfully reproduced, the model is too specific and lacks generalization power. As a result, predictions for future data are generally of lower quality than expected. And as for the interpretability of the model: this is very much dependent on the composition of the training set. A small change in the data can lead to a completely different tree. As we will see, this is a disadvantage that can be turned into an advantage: combinations of tree-based classifiers often give stable and accurate predictions. These so-called Random Forests, taking away many of the disadvantages of simple tree-based classifiers while keeping the good characteristics, enjoy huge popularity and will be treated in Sect. 9.7.2.

7.4 More Complicated Techniques

When relatively simple models like LDA or KNN do not succeed in producing models with good predictive capabilities, one can ask the question: why do we fail? Is it because the data just do not contain enough information to build a useful model? Or are the models we have tried too simple? Do we need something more flexible, perhaps nonlinear? The distinction between information-poor data and complicated class boundaries is often hard to make.

In this section, we will treat two popular nonlinear techniques from the domain of Machine Learning with complementary characteristics: whereas *Support Vector Machines* (SVMs) are very useful when the number of objects is not too large, *Artificial Neural Networks* (ANNs) should only be applied when there are ample training cases available. Conversely, SVMs are applicable in high-dimensional cases whereas ANNs are not: very often, a data reduction step like PCA is employed to bring the number of variables down to a manageable size. These two techniques do share one important property: they are very flexible indeed, and capable of modelling the most complex relationships. This puts a large responsibility on the researcher for thorough validation, especially since there are several parameters to tune. Because the theory behind the methods is rather extensive, we will only sketch the contours—interested readers are referred to the literature for more details.

Fig. 7.11 The basic idea behind SVM classification: the separating hyperplane (here, in two dimensions, a line) is chosen in such a way that the margin is maximal. Points on the margins (the dashed lines) are called "support vectors". Clearly, the margins for the separating line with slope 2/3 are much further apart than for the vertical boundary

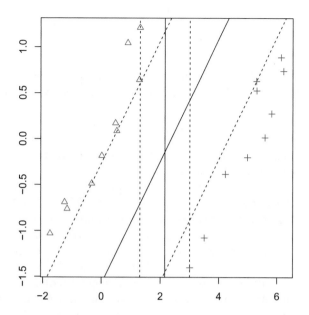

7.4.1 Support Vector Machines

SVMs (Vapnik 1995; Cristianini and Shawe-Taylor 2000; Schölkopf and Smola 2002) in essence are binary classifiers, able to discriminate between two classes. They aim at finding a separating hyperplane maximizing the distance between the two classes. This distance is called the *margin* in SVM jargon; a synthetic example, present in almost all introductions to SVMs, is shown in Fig. 7.11. Although both classifiers, indicated by the solid lines, perfectly separate the two classes, the classifier with slope 2/3 achieves a much bigger margin than the vertical line. The points that are closest to the hyperplane are said to lie on the margins, and are called *support vectors*—these are the only points that matter in the classification process itself. Note however that all other points have been used in setting up the model, i.e., in determining which points are support vectors in the first place. The fact that only a limited number of points is used in the predictions for new data is called the *sparseness* of the model, an attractive property in that it focuses attention to the region that matters, the boundary between the classes, and ignores the exact positions of points far from the battlefield.

More formally, a separating hyperplane can be written as

$$\boldsymbol{w}\boldsymbol{x} - b = 0 \tag{7.20}$$

The margin is the distance between two parallel hyperplanes with equations

$$\boldsymbol{wx} - b = -1 \qquad\qquad (7.21)$$
$$\boldsymbol{wx} - b = 1 \qquad\qquad (7.22)$$

and is given by $2/\|\boldsymbol{w}\|$. Therefore, maximizing the margin comes down to minimizing $\|\boldsymbol{w}\|$, subject to the constraint that no data points fall within the margin:

$$c_i(\boldsymbol{wx}_i) \leq 1 \qquad\qquad (7.23)$$

where c_i is either -1 or 1, depending on the class label. This is a standard quadratic programming problem.

It can be shown that these equations can be rewritten completely in terms of inner products of the support vectors. This so-called *dual representation* has the big advantage that the original dimensionality of the data is no longer of interest: it does not really matter whether we are analyzing a data matrix with two columns, or a data matrix with ten thousand columns. By applying suitable *kernel functions*, one can transform the data, effectively leading to a representation in higher-dimensional space. Often, a simple discrimination function can be obtained in this high-dimensional space, which translates into an often complex class boundary in the original space. Because of the dual representation, one does not need to know the exact transformation—it suffices to know that it exists, which is guaranteed by the use of kernel functions with specific properties. Examples of suitable kernels are the polynomial and gaussian kernels. More details can be found in the literature (e.g., Hastie et al. 2001).

Package **e1071** provides an interface to the LIBSVM library[3] through the function svm. Autoscaling is applied by default. Modelling the Barbera and Grignolino classes leads to the following results:

```
> wns.df <-
+    data.frame(vint = vnt,
+               flavonoids = wns[,"flavonoids"],
+               proline = wns[,"proline"])
> wns.svm <- svm(vint ~ ., data = wns.df[wines.odd2, ])
> wns.svmpred <- predict(wns.svm, wns.df[wines.even2, ])
> table(wns.df$vint[wines.even2], wns.svmpred)
             wns.svmpred
             Barbera Grignolino
  Barbera         22          2
  Grignolino       4         31
```

These default settings lead to a reasonable of the test set.

One attractive feature of SVMs is that they are able to handle fat data matrices (where the number of features is much larger than the number of objects) without any problem. Let us see, for instance, how the standard SVM performs on the prostate data. We will separate the cancer samples from the other control class—again, we are considering only the first 1000 variables. Using the cross = 10 argument, we

---

[3] See http://www.csie.ntu.edu.tw/~cjlin/libsvm/.

perform ten-fold crossvalidation, which should give us some idea of the performance
on the test set:

```
> prost.svm <- svm(type ~ ., data = prost.df, subset = prost.odd,
+                   cross = 10)
> summary(prost.svm)

Call:
svm(formula = type ~ ., data = prost.df, cross = 10, subset = prost.odd)


Parameters:
   SVM-Type:  C-classification
 SVM-Kernel:  radial
       cost:  1
      gamma:  0.001

Number of Support Vectors:  88

 ( 38 50 )


Number of Classes:  2

Levels:
 control pca

10-fold cross-validation on training data:

Total Accuracy: 92
Single Accuracies:
 83.333 84.615 100 92.308 100 84.615 100 100 91.667 84.615
```

This summary shows us that rather than the complete training set of 125 samples,
only 88 are seen as support vectors (for SVMs already quite a large fraction). The
prediction accuracies for the left out segments vary from 83 to 83%, with an overall
error estimate of 92%. Let us see whether the test set can be predicted well:

```
> prost.svmpred <- predict(prost.svm, newdata = prost.df[prost.even,])
> table(prost.type[prost.even], prost.svmpred)
          prost.svmpred
           control pca
  control     33    7
  pca          1   83
```

Six misclassifications out of 124 cases, nicely in line with the crossvalidation error
estimate, is better than anything we have seen so far—not a bad result for default
settings.

7.4.1.1 Extensions to More than Two Classes

The fact that only two-class situations can be tackled by basic forms of SVMs is a severe limitation: in reality, it often happens that we should discriminate between several classes. The standard approach is to turn one multi-class problem into several two-class problems. More specifically, one can perform one-against-one testing, where every combination of single classes is assessed, or one-against-all testing. In the latter case, the question is rephrased as: "to be class A or not to be class A"—the advantage is that, in the case of n classes, only n comparisons need to be made, whereas in the one-against-one case $n(n-1)/2$ models must be fitted. The disadvantage is that the class boundaries may be much more complicated: class "not A" may be very irregular in shape. The default in the function svm is to assess all one-against-one classifications, and use a voting scheme to pinpoint the final winning class.

To show how this works we again concentrate on two dimensions only so that we can visualize the results. First we set up the SVM model using the odd-numbered rows only:

```
> wines.svm <- svm(vint ~ flavonoids + proline, data = wines.df,
+                   subset = wines.odd)
> wines.svmpred.trn <- predict(wines.svm)
> wines.svmpred.tst <-
+    predict(wines.svm, newdata = wines.df[wines.even, ])
> sum(wines.svmpred.trn == vint.trn) / length(wines.odd)
[1] 0.91011
> sum(wines.svmpred.tst == vint.tst) / length(wines.even)
[1] 0.90909
```

Predictions are very good, both for the training data (the odd rows of the data frame) and the test data (the even rows). Next, we can plot the class boundaries, and project the values of the test data on top to get a visual impression:

```
> plot(wines.svm, wines.df[wines.even, ], proline ~ flavonoids,
+      color.palette = softbrg)
```

This code leads to the left plot in Fig. 7.12. The background colours, indicate the predicted class for each region in the plot. They are obtained in a way very similar to the code used to produce the contour lines in Fig. 7.3 and similar plots. Plotting symbols show the positions of the support vectors—these are shown as crosses, whereas regular data points, unimportant for this SVM model, are shown as open circles. The 8 misclassifications can easily be spotted in the figure.

7.4.1.2 Finding the Right Parameters

The biggest disadvantage of SVMs is the large number of tuning parameters. One should choose an appropriate kernel, and, depending on this kernel, values for two

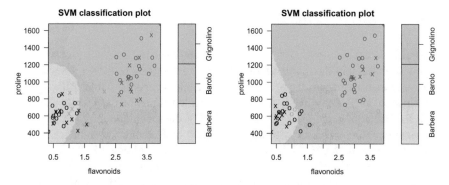

Fig. 7.12 SVM classification plots for the two-dimensional wine data (training data only). Support vectors are indicated by crosses; regular data points by open circles. Left plot: default settings of svm. Right plot: best SVM model with a polynomial kernel, obtained with best.svm

or three parameters. A special convenience function, tune, is available in the **e1071** package, which, given a choice of kernel, varies the settings over a grid, calculates validation values such as crossvalidated prediction errors, and returns an object of class tune containing all validation results. A related function is best which returns the model with the best validation performance. If we wanted to find the optimal settings for the three parameters coef0, gamma and cost using a polynomial kernel (the default kernel is a radial basis function), we could do it like this:

```
> set.seed(7)
> wines.bestsvm <-
+    best.svm(vint ~ flavonoids + proline, data = wines.df,
+             kernel = "polynomial",
+             coef0 = seq(-.5, .5, by = .1),
+             gamma = 2^(-1:1), cost = 2^(2:4))
```

The predictions with these settings then lead to the following results:

```
> wines.bestsvmpred.trn <-
+    predict(wines.bestsvm, newdata = wines.df[wines.odd, ])
> wines.bestsvmpred.tst <-
+    predict(wines.bestsvm, newdata = wines.df[wines.even, ])
> sum(wines.bestsvmpred.trn == vint.trn) / length(vint.trn)
[1] 0.92135
> sum(wines.bestsvmpred.tst == vint.tst) / length(vint.tst)
[1] 0.92045
```

For both the training and test data, one fewer misclassification error is made; however, the classification plot, shown in the right of Fig. 7.12 looks quite different from the earlier version. The differences in areas where no samples are present may seem not particularly interesting—however, they may become very relevant when new samples are classified. Note that also the number and position of support vectors is quite different.

7.4.2 *Artificial Neural Networks*

Artificial Neural Networks (ANNs, also shortened to neural networks, NNs) form a
family of extremely flexible modelling techniques, loosely based on the way neurons in human brains are thought to be connected—hence the name. Although the
principles of NNs had already been defined in the fifties of the previous century
with Rosenblatt's perceptron (Rosenblatt 1962), the technique only really caught
on some twenty years later with the publication of Rumelhart's and McClellands
book (Rumelhard and McClelland 1986). Many different kinds of NNs have been
proposed; here, we will only treat the flavor that has become known as *feed-forward
neural networks*, *backpropagation networks*, after the name of the training rule (see
below), or *multi-layer perceptrons*.

Such a network consists of a number of units, typically organized in three layers,
as shown in Fig. 7.13. When presented with input signals s_i, a unit will give an output
signal s_o corresponding to a transformation of the sum of the inputs:

$$s_o = f\left(\sum_i s_i\right) \tag{7.24}$$

For the units in the input layer, the transformation is usually the identity function,
but for the middle layer (the *hidden layer* typically sigmoid transfer functions or

Fig. 7.13 The structure of a
feedforward NN with three
input units, four hidden
units, two bias units and two
output units

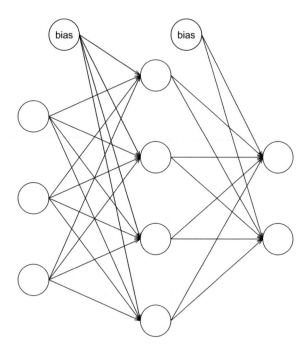

threshold functions are used.   For the hidden and output layers, special *bias units* are traditionally added, always having an output signal of +1 (Ripley 1996). Network structure is very flexible. It is possible to use multiple hidden layers, remove links between specific units, to add connections skipping layers, or even to create feedback loops where output is again fed to special input units. However, the most common structure is to have a fully connected network such as the one depicted in Fig. 7.13, consisting of one input layer, one hidden layer and one output layer. One can show that adding more hidden layers will not lead to better predictions (although in some cases it is reported to speed up training). Whereas the numbers of units in the input and output layers are determined by the data, the number of units in the hidden layer is a parameter that must be optimized by the user.

Connections between units are weighted: an output signal from a particular unit is sent to all connected units in the next layer, multiplied by the respective weights. These weights, in fact form the model for a particular network topology—training the network comes down to finding the set of weights that gives optimal predictions. The most popular training algorithm is based on a steepest-descent based adaption of the weights upon repeated presentation of training data. The gradient is determined by what is called the *backpropagation* rule, a simple application of the chain rule in obtaining derivatives. Many other training algorithms have been proposed as well.

In R, several packages are available providing general neural network capabilities, such as **AMORE** and **neuralnet** (Günther and Fritsch 2010). We will use the **nnet** package, one of the recommended R packages, featuring feed-forward networks with one hidden layer, several transfer functions and possibly skip-layer connections. It does not employ the usual backpropagation training rule but rather the optimization method provided by the R function `optim`. The target values, here the labels of the vintages, have to be presented as a membership matrix, here containing three columns, one for each type of wine. Each row contains 1 at the correct label of the sample, and zeros in the other two positions. The conversion of a factor to a membership matrix is done by the `classvec2classmat` function from the **kohonen** package—below, the first three lines of the membership matrix (all Barolos) are shown:

```
> membership.trn <- classvec2classmat(vint.trn)
> head(membership.trn, 3)
     Barbera Barolo Grignolino
[1,]       0      1          0
[2,]       0      1          0
[3,]       0      1          0
```

For the (autoscaled) training set of the wine data, the network is trained as follows:

```
> wines.nnet <- nnet(x = wines.trn.sc,
+                    y = membership.trn,
+                    size = 4)
# weights:   71
initial  value 64.600576
iter  10 value 28.923577
```

```
iter   20 value 0.012965
iter   30 value 0.002048
iter   40 value 0.000696
iter   50 value 0.000530
final   value 0.000096
converged
```

Although the autoscaling is not absolutely necessary (the same effect can be reached by using different weights for the connections of the input units to the hidden layer) it does make it easier for the network to reach a good solution—without autoscaling the data, the optimization easily gets stuck in a local optimum. Here convergence is reached very quickly. In practice, multiple training sessions should be performed, and the one with the smallest (crossvalidated) training error should be selected. An alternative is to use a (weighted) prediction using all trained networks.

As expected for such a flexible fitting technique, the training data are reproduced perfectly:

```
> membership.pred <- predict(wines.nnet)
> training.pred <- classmat2classvec(membership.pred)
> table(vint.trn, training.pred)
              training.pred
vint.trn     Barbera Barolo Grignolino
   Barbera        24      0          0
   Barolo          0     29          0
   Grignolino      0      0         36
```

Luckily, also the test data are predicted very well here:

```
> table(vint.tst,
+        classmat2classvec(predict(wines.nnet, wines.tst.sc)))

vint.tst     Barbera Barolo Grignolino
   Barbera        24      0          0
   Barolo          0     29          0
   Grignolino      1      1         33
```

Several default choices have been made under the hood of the nnet function: the type of transfer functions in the hidden layer and in the output layer, the number of iterations, whether least-squares fitting or maximum likelihood fitting is done (default is least-squares), and several others. The only explicit setting in this example is the number of units in the hidden layer, and this immediately is the most important parameter, too. Choosing too many units will lead to a good fit of the training data but potentially bad generalization—overfitting. Too few hidden units will lead to a model that is not flexible enough.

A convenience function tune.nnet is available in package **e1071**, similar to tune.svm. This will test neural networks of a specific architecture a number of times (the default is five) and collect measures of predictive performance (obtained by either crossvalidation or bootstrapping, see Chap. 9). Let us see whether our

Fig. 7.14 Tuning neural networks: selecting the optimal number of nodes in the hidden layer

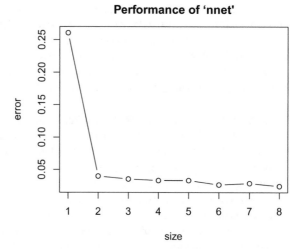

(arbitrary) choice of four hidden units can be improved upon, now using the formula interface of the nnet function:

```
> wines.trn.sc.df <- data.frame(vintage = vint.trn, wines.trn.sc)
> (wines.nnetmodels <-
+    tune.nnet(vintage ~ ., data = wines.trn.sc.df,
+              size = 1:8, trace = FALSE))
```

Generic summary and a plot methods are available—for the corresponding plot, see Fig. 7.14. Clearly, one hidden unit is not enough, and two hidden units are not much worse than four, or even eight (although changing the y scale could make us rethink that statement). Instead of using tune.nnet, one can also apply best.nnet—this function directly returns the trained model with the optimal parameter settings:

```
> best.wines.nnet <-
+    best.nnet(vintage ~ ., data = wines.trn.sc.df,
+              size = 1:8, trace = FALSE)
> table(vint.tst,
+       predict(best.wines.nnet,
+               newdata = data.frame(wines.tst.sc),
+               type = "class"))

vint.tst      Barbera Barolo Grignolino
   Barbera        24      0          0
   Barolo          0     29          0
   Grignolino      4      0         31
```

Indeed, we see one fewer misclassification. Note that here, "optimal" simply means the network with the lowest crossvalidation error. However, this may be too optimistic: especially with very flexible models like ANNs and the tree-based methods we saw earlier, overfitting is a real danger. The idea is that too complex models have enough flexibility to learn data "by heart" whereas models of the right complexity are forced to focus on more general principles. One rule of thumb is to use the simplest possible model that is not worse than the best model to be as conservative as possible. In this case, one would expect a network with only two hidden neurons to perform better for new, unseen, data than a network with four or eight. We will come back to this in Chap. 9.

In the example above we used the default stopping criterion of the nnet function, which is to perform 100 iterations of complete presentations of the training data. In several publications, scientists have advocated continuously monitoring prediction errors throughout the training iterations, in order to prevent the network from overfitting. In this approach, training should be stopped as soon as the error of the validation set starts increasing. Apart from the above-mentioned training parameters, this presents an extra level of difficulty which becomes all the more acute with small data sets. To keep these problems manageable, one should be very careful in applying neural networks in situations with few cases; the more examples, the better.

7.4.2.1 Deep Learning

Since 2010, a novel development in neural networks called Deep Learning (DL, Goodfellow et al. 2016) has taken center stage with applications in areas like computer vision (Uijlings et al. 2013; Gatys et al. 2016; Badrinarayanan et al. 2017), speech recognition (Hinton et al. 2012; Deng et al. 2013; Nassif et al. 2019) and many others. At that point in time, developments in GPUs, graphics processing units allowing massively parallel computations coincided with easy access to large data sets such as ImageNet (Russakovsky et al. 2015), a collection of millions of annotated images. Open-source software was available, too, and the interest of companies like Google and Microsoft made sure large steps were made. Today, many of the top-performing approaches in difficult benchmark problems are based on Deep Learning.

So what is different, compared to the neural networks in the previous sections? From a structural viewpoint, not that much. Just like the neural networks from the nineties can be seen as perceptrons stitched together in a particular structure, DL networks can be described as collections of neural networks such as the ones in Fig. 7.13. What *is* new is that many more layers are used (DL networks with more than one hundred layers are no exception) and that layers are included with a purpose: in image processing applications we typically see, amongst others, convolutional layers and pooling layers, applied in alternating fashion. Each of these layers serves a particular purpose—subsequent layers are not fully connected like in the network of Fig. 7.13 but only connected if there is a reason for it. In this way, the DL network is able to aggregate the raw input data into more and more abstract features that eventually will be combined to obtain the final answer. The increased amount of structure within the DL net restricts the number of weights that need to be optimized. Regularization methods (see, e.g., Sects. 8.4 and 10.2) are employed routinely in order to keep the weights small and prevent overfitting (Efron and Hastie 2016), effectively removing the number of training iterations as a parameter to be optimized: more is better in these cases. Still, the training is a daunting task: many weights are optimized, and for this a large (large!) number of training examples needs to be provided.

A major hurdle for classification applications is that in almost all cases the training examples need to be annotated, i.e., the ground truth needs to be known. Modern sensing devices like cameras have no problems in generating terabytes and more of data, but what the true class of the image is still needs to be decided, an area that is being exploited commercially nowadays. Chemometrics is typically concerned with data characterized by multivariate responses, recorded for relatively few samples, so DL applications are still rare but they will certainly come.

Chapter 8
Multivariate Regression

In Chaps. 6 and 7 we have concentrated on finding groups in data, or, given a grouping, creating a predictive model for new data. The last situation is "supervised" in the sense that we use a set of examples with known class labels, the training set, to build the model. In this chapter we will do something similar—now we are not predicting a discrete class property but rather a continuous variable. Put differently: given a set of independent real-valued variables (matrix X), we want to build a model that allows prediction of Y, consisting of one, or possibly more, real-valued dependent variables. As in almost all regression cases, we here assume that errors, normally distributed with constant variance, are only present in the dependent variables, or at least are so much larger in the dependent variables that errors in the independent variables can be ignored. Of course, we also would like to have an estimate of the expected error in predictions for future data.

8.1 Multiple Regression

The usual multiple least-squares regression (MLR), taught in almost all statistics courses, is modelling the relationship

$$Y = XB + \mathcal{E} \tag{8.1}$$

where B is the matrix of regression coefficients and $\mathcal{E}$ contains the residuals. The regression coefficients are obtained by

$$B = (X^T X)^{-1} X^T Y \tag{8.2}$$

with variance-covariance matrix

© Springer-Verlag GmbH Germany, part of Springer Nature 2020
R. Wehrens, *Chemometrics with R*, Use R!,
https://doi.org/10.1007/978-3-662-62027-4_8

$$\text{Var}(\boldsymbol{B}) = (\boldsymbol{X}^T \boldsymbol{X})^{-1} \sigma^2 \tag{8.3}$$

The residual variance σ^2 is typically estimated by

$$\hat{\sigma}^2 = \frac{1}{n - p - 1} \sum_{i=1}^{n} (y_i - \hat{y}_i)^2 \tag{8.4}$$

MLR has a number of attractive features, the most important of which is that it is the Best Linear Unbiased Estimator (BLUE) when the assumption of uncorrelated normally distributed noise with constant variance is met (Mardia et al. 1979). The standard deviations of the individual coefficients, given by the square roots of the diagonal elements of the variance-covariance matrix $\text{Var}(\boldsymbol{B})$, can be used for statistical testing: variables whose coefficients are not significantly different from zero are sometimes removed from the model. Since removing one such variable will in general lead to different estimates for the remaining variables, this results in a stepwise variable selection approach (see Chap. 10).

For the odd-numbered samples in the `gasoline` data, a regression using four of the 401 wavelengths, evenly spaced over the entire range to reduce correlations between the wavelengths would yield

```
> X <- gasoline$NIR[, 100*(1:4)]
> Y <- gasoline$octane
> Xtr <- cbind(1, X[gas.odd, ])
> Ytr <- Y[gas.odd]
> Bs <- t(solve(crossprod(Xtr), t(Xtr)) %*% Ytr)
> Bs
          1098 nm 1298 nm 1498 nm 1698 nm
[1,] 64.482   1312.9 -1607.5  229.94  26.244
```

Adding the column of ones using the `cbind` function on the fifth line in the example above causes an intercept to be fitted as well. The `solve` statement is the direct implementation of Eq. 8.2. This also works when Y is multivariate—the regression matrix $\boldsymbol{B}$ will have one column for every variable to be predicted.

Rather than using this explicit matrix inversion, one would use the standard linear model function `lm`, which also provides the usual printing, plotting and summary functions. The `lm` function is provided with a `data.frame` as input. To avoid funny-looking names we'll make a temporary `data.frame`:

```
> gas.df <- data.frame(octane = gasoline$octane,
+                      V100 = gasoline$NIR[, 100],
+                      V200 = gasoline$NIR[, 200],
+                      V300 = gasoline$NIR[, 300],
+                      V400 = gasoline$NIR[, 400])
> Blm <- lm(octane ~ ., data = gas.df)
> summary(Blm)

Call:
lm(formula = octane ~ ., data = gas.df)

Residuals:
    Min      1Q Median      3Q     Max
 -4.328  -0.874  0.383   0.811   2.275

Coefficients:
             Estimate Std. Error t value Pr(>|t|)
(Intercept)     72.59      11.52    6.30 5.2e-08 ***
V100           714.94     178.39    4.01 0.00019 ***
V200          -957.29     224.92   -4.26 8.2e-05 ***
V300           154.76      66.29    2.33 0.02324 *
V400            12.01       7.24    1.66 0.10260
---
Signif. codes:  0 '***' 0.001 '**' 0.01 '*' 0.05 '.' 0.1 ' ' 1

Residual standard error: 1.35 on 55 degrees of freedom
Multiple R-squared:  0.278, Adjusted R-squared:  0.226
F-statistic:  5.3 on 4 and 55 DF,  p-value: 0.0011
```

The `lm` function automatically fits an intercept; there is no need to explicitly add a column of ones to the matrix of independent variables. Under the usual assumption of normal independent and identically distributed (*iid*) residuals, the p-values for the coefficients are gathered in the last column: all but one coefficients are significant at the $\alpha = 0.05$ level.

8.1.1 Limits of Multiple Regression

Unfortunately, however, there are some drawbacks. In the context of the natural sciences, the most important perhaps is the sensitivity to correlation in the independent variables. This can be illustrated using the following example. Suppose we have a model that looks like this:

$$y = 2 + x_1 + 0.5x_2 - 2x_3$$

and further suppose that x_2 and x_3 are highly correlated ($r \approx 1.0$). This means that any of the following models will give more or less the same predictions:

$$y = 2 + x_1 - 1.5x_2$$
$$y = 2 + x_1 - 1.5x_3$$
$$y = 2 + x_1 + 5.5x_2 - 7x_3$$
$$y = 2 + x_1 + 1000.5x_2 - 1002x_3$$

So what is so bad about that? If all x values would exactly be the same, not even that much, but in practice there will be errors, affecting all these models in different ways. Especially when coefficients are large, differences can be appreciable, one reason to prefer, from a set of equivalent models, the one with the smallest regression coefficients. Furthermore, confidence intervals for the regression coefficients are based on the assumption of independence, which clearly is violated in this case: any coefficient value for x_2 can be compensated for by x_3, and variances for the x_2 and x_3 coefficients will be infinite. Also in cases where there is less than perfect correlation, we will see more unstable models, in the sense that the variances of the coefficient estimates will get large and predictions less reliable.

To be fair, ordinary multiple regression will not allow you to calculate the model in pathological cases like the above: matrix $X^T X$ will be singular, indicating that infinitely many inverse matrices, and, conversely, many different coefficient vectors, are possible—it is comparable to choosing the best straight line through only one point. Another case where the inverse cannot be calculated is the situation where there are more independent variables than samples, which indeed is the case for the `gasoline` data:

```
> Xtr <- cbind(1, gasoline$NIR[gas.odd, ])
> solve(crossprod(Xtr), t(Xtr)) %*% Ytr
```

This will throw an error, which was the primary reason to select four of the variables in the gasoline example in the beginning of this section. Unfortunately, in almost all applications of spectroscopy the number of variables far exceeds the number of samples; the correlations between variables is often high, too.

One possibility to tackle the above problem is to calculate a *pseudoinverse* matrix X^+—such a pseudoinverse, or a generalized inverse, has the property that

$$X X^+ = 1 \tag{8.5}$$

and can be applied to non-square matrices as well as square matrices. The most often used variant is   the Moore-Penrose inverse, available in R as function `ginv` in package **MASS**:

```
> Blm <- ginv(Xtr) %*% Ytr
```

The Moore-Penrose inverse uses the singular value decomposition of the data matrix:

$$X^{-1} = \left(U D V^T\right)^{-1} = V D^{-1} U^T \tag{8.6}$$

The trick is to ignore the singular values in D that are zero—in the inverse matrix D^{-1} these will still have a value of zero and the corresponding rows in V and U will be disregarded. In practice, of course, a threshold will have to be used which is usually taken to be dependent on the machine precision. Values smaller than the threshold will be set to zero in the inverse of D. Singular values which are slightly larger, however, may exert a large influence on the result, and in many cases the generalized inverse is not very stable.

It is taking this idea one step further to restrict the number of singular values in Eq. 8.6 to only the most important principal components. This is the basis of Principal Component Regression (PCR). PCR basically performs a regression on the *scores* of X, where a suitable number of latent variables has to be chosen. An alternative, Partial Least Squares (PLS) regression employs the same basic idea, but takes the dependent variable into account when defining scores and loadings, whereas PCR concentrates on capturing variance in X only. In both techniques, the inversion of the covariance matrix is simple because of the orthogonality of the scores. The price we pay is threefold: vital information may be lost because of the data compression, we have to choose the degree of compression, i.e., the number of latent variables, and finally it is no longer possible to derive analytical expressions for the prediction error and the variances of individual regression coefficients. To be able to say something about the optimal number of latent variables, and about the expected error of prediction, crossvalidation or similar techniques must be used (see Chap. 9).

We have already seen that in general models with small regression coefficients are to be preferred. It can be shown that PLS as well as PCR actually shrink the regression coefficients towards zero (Hastie et al. 2001). They are therefore biased methods: the coefficients on average will be smaller in absolute value than the unknown, "true", coefficients—however, this will be compensated for by a much lower variance. Other approaches, based on explicit penalization of the regression coefficients, can be used as well. If a quadratic (L_2) penalty is employed, the result is called *ridge regression*. A penalty in the form of absolute values (an L_1 penalty) leads to the *lasso*, whereas a combination of L_1 and L_2 penalties is known as the *elastic net*. Methods based on the L_1 norm have the advantage that many of the coefficients will have a value of zero, thereby implicitly performing variable selection. They will be treated, along with explicit variable selection methods, in Chap. 10.

8.2 PCR

The prime idea of PCR is to use scores rather than the original data for the regression step. This has two advantages: scores are orthogonal, so there are no problems with correlated variables, and secondly, the number of PCs taken into account usually is much lower than the number of original variables. This considerably reduces the number of coefficients that must be estimated, which in turn leads to more degrees of freedom for the estimation of errors. Of course, we have the added problem that we have to estimate how many PCs to retain.

8.2.1 The Algorithm

For the moment, let us select a PCs; matrices T, P, etcetera, will have a columns. The regression model is then built using the *reconstructed* matrix $\tilde{X} = T P^T$ rather than the original matrix:

$$Y = \tilde{X} B + \mathcal{E} = T(P^T B) + \mathcal{E} = T A + \mathcal{E} \tag{8.7}$$
$$A = (T^T T)^{-1} T^T Y \tag{8.8}$$

where $A = P^T B$ will contain the regression coefficients for the scores. The crossproduct matrix of T is diagonal (remember, T is orthogonal) so can be easily inverted. The regression coefficients for the scores can be back-transformed to coefficients for the original variables:

$$\begin{aligned} B &= P A \\ &= P(T^T T)^{-1} T^T Y \end{aligned} \tag{8.9}$$

This can be simplified further by resubstituting $T = U D$ (from the SVD routine):

$$\begin{aligned} B &= P(D U^T U D)^{-1} D U^T Y \\ &= P D^{-2} D U^T Y \\ &= P D^{-1} U^T Y \end{aligned} \tag{8.10}$$

In practice, one always performs PCR on a mean-centered data matrix. In Chap. 4 we have seen that without mean-centering the first PC often dominates and is very close to the vector of column means, an undesirable situation. By mean-centering, we explicitly force a regression model without an intercept. The result is that the coefficient vector B does not contain an abscissa vector b_0; it should be calculated explicitly by taking the difference between the mean y values and the mean *predicted* y-values.

$$b_0 = \bar{y} - X B \tag{8.11}$$

For every variable in Y, we will find one number in b_0.

Let's see how this works for the gasoline data. We will model the gas.odd rows, now based on the complete NIR spectra. We start by mean-centering the data, based only on the gas.odd rows:

```
> X <- scale(gasoline$NIR, scale = FALSE,
+            center = colMeans(gasoline$NIR[gas.odd, ]))
```

Note that we do not use autoscaling, since that would blow up noise in uninformative variables. Next, we calculate scores and use these as independent variables in a regression. For the moment, we choose five PCs.

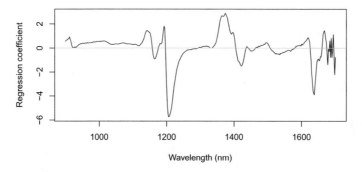

Fig. 8.1 Regression coefficients for the gasoline data (based on the gas.odd rows), obtained by PCR using five PCs

```
> Xgas.odd.svd <- svd(X[gas.odd, ])
> Xgas.odd.scores <- Xgas.odd.svd$u %*% diag(Xgas.odd.svd$d)
> gas.odd.pcr <-
+    lm(gasoline$octane[gas.odd] ~ I(Xgas.odd.scores[, 1:5]) - 1)
```

The - 1 in the formula definition prevents lm from fitting an intercept; the other coefficients are not affected, whether an intercept is fitted or not, but removing it makes the comparison below slightly easier. The regression coefficients of the wavelengths, the original variables, are obtained by multiplying the regression coefficients of the scores with the corresponding loadings:

```
> gas.odd.coefs <- coef(gas.odd.pcr) %*% t(Xgas.odd.svd$v[, 1:5])
```

The same model can be produced by the pcr function from the **pls** package (Mevik and Wehrens 2007):

```
> gasoline.pcr <- pcr(octane ~ ., data = gasoline,
+                     subset = gas.odd, ncomp = 5)
> all.equal(c(coef(gasoline.pcr)), c(gas.odd.coefs))
> plot(wavelengths, coef(gasoline.pcr), type = "l",
+      xlab = "Wavelength (nm)", ylab = "Regression coefficient")
> abline(h = 0, col = "gray")
```

The last line produces the plot of the regression coefficients, shown in Fig. 8.1. As usual, the intercept is not visualized. One can clearly see features such as the peaks around 1200 and 1400 nm. Often, such important variables can be related to physical or chemical phenomena.

The model can be summarized by the generic function summary.mvr:

```
> summary(gasoline.pcr)
Data:    X dimension: 30 401
Y dimension: 30 1
Fit method: svdpc
Number of components considered: 5
TRAINING: % variance explained
           1 comps   2 comps   3 comps   4 comps   5 comps
X            74.318     86.26     91.66     96.11     97.32
octane        9.343     11.32     16.98     97.22     97.26
```

Clearly, the first component focuses completely on explaining variation in X; it is the fourth component that seems most useful in predicting Y, the octane number.

8.2.2   Selecting the Optimal Number of Components

How much variance of Y is explained is one criterion one could use to determine the optimal number of PCs. More often, however, one monitors the (equivalent) root-mean-square error (RMS or RMSE):

$$\text{RMS} = \sqrt{\frac{\sum_i^n (\hat{y}_i - y_i)^2}{n}} \qquad (8.12)$$

where $\hat{y}_i - y_i$ is the difference between predicted and true value, and n is the number of predictions. A simple R function to find RMS values is the following:

```
> rms <- function(x, y) sqrt(mean((x-y)^2))
```

It is important to realize that both criteria, the amount of variance explained as well as the RMSE, assess the fit of the model to the training data, i.e., the quality of the reproduction rather than the predictive abilities of the model.

The **pls** package comes with an extractor function for RMS estimates:

```
> RMSEP(gasoline.pcr, estimate = "train", intercept = FALSE)
1 comps   2 comps   3 comps   4 comps   5 comps
  1.3467    1.3319    1.2887    0.2358    0.2343
```

The intercept = FALSE argument prevents the RMS error based on nothing but the average of the dependent variable to be printed. These numbers are cumulative, so the number under "4 comps" signifies the error when using components 1–4. The error when *only* using the fourth PC in the regression is given by

```
> RMSEP(gasoline.pcr, estimate = "train", comp = 4)
[1]   0.6287
```

It is only half the size of the error of prediction using PCs one to three, showing that the fourth PC is very important in the prediction model. Nevertheless, the combination of the fourth with the first three components leads to an even better model. Adding the fifth does not seem worthwhile.

One should be very cautious in interpreting these numbers—reproduction of the training set is not a reliable way to assess prediction accuracy, which is what we really are after. For one thing, adding another component will (by definition) always decrease the error. Predictions for new data, however, may not be as good: the model often is too focused on the training data, the phenomenon known as overfitting that we also saw in Sect. 7.3.1.

A major difficulty in PCR modelling is therefore the choice of the optimal number of PCs to retain. Usually, crossvalidation is employed to estimate this number: if the number of PCs is too small, we will incorporate too little information, and our model will not be very useful—the crossvalidation error will be high. If the number of PCs is too large, the training set will be reproduced very well, but again, the crossvalidation error will be large because the model is not able to provide good predictions for the left-out samples. So the strategy is simple: we perform the regression for several numbers of PCs, calculate the crossvalidation errors, put the results in a graph, *et voilà*, we can pick the number we like best.

In LOO crossvalidation, n different models are created, each time omitting another sample y_{-i} from the training data. Equation 8.12 then is used to calculate an RMS estimate. Alternatively, a group of samples is left out simultaneously. One often sees the name RMSCV or RMSECV, to indicate that it is the RMS value derived from a crossvalidation procedure. Of course, this procedure can be used in other contexts as well; the RMSEP usually is associated with prediction of unseen test data, RMSEV is an error estimate from some form of validation, and RMSEC is the calibration error, indicating how well the model fits the training data. The RMSEC value is in almost all cases the lowest; it measures how well the model represents the data on which it is based. The RMSEP is an estimate of the thing that really interests us, the error of prediction, but it can only reliably be calculated directly when large training and test sets are available and the training data are representative for the test data. In practice, RMSE(C)V estimates are the best we can do.

In the **pls** package, the function RMSEP is used to calculate all these quantities—it takes an argument estimate which can take the values "train" (used in the example above), "CV" and "test" for calibration, crossvalidation and test set estimates, respectively. The pcr function has an optional argument for crossvalidation: validation can be either "none", "LOO", or "CV". In the latter case, 10 segments are used in the crossvalidation by default ("leave 10% out"). Application to the gasoline data leads to the following results:

```
> gasoline.pcr <- pcr(octane ~ ., data = gasoline, subset = gas.odd,
+                     validation = "LOO", ncomp = 10)
> plot(gasoline.pcr, "validation", estimate = "CV")
```

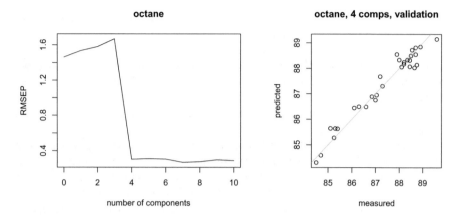

Fig. 8.2 Validation plot for PCR regression on the gasoline data (left) and prediction quality of the optimal model, containing four PCs (right)

This leads to the validation plot in the left panel of Fig. 8.2, showing the RMSECV estimate[1] against the number of PCs. Not unexpectedly, four PCs clearly are a very good compromise between model complexity and predictive power. For the argument `estimate`, one can also choose `"adjCV"`, which is a bias-corrected error estimate (Mevik and Cederkvist 2004).

Zooming in on the plot would show that the absolute minimum is at seven PCs, and perhaps taking more than ten PCs into consideration would lead to an lower global minimum. However, one should keep in mind that an RMS value is just an estimate with an associated variance, and differences are not always significant. Moreover, the chance of overfitting increases with a higher number of components. Numerical values are accessible in the `validation` list element of the fitted object. The RMS values plotted in Fig. 8.2 can be assessed as follows:

```
> RMSEP(gasoline.pcr, estimate = "CV")
(Intercept)      1 comps      2 comps      3 comps      4 comps
     1.4631       1.5351       1.5802       1.6682       0.3010
    5 comps      6 comps      7 comps      8 comps      9 comps
     0.3082       0.3031       0.2661       0.2738       0.2949
   10 comps
     0.2861
```

which is the same as

```
> sqrt(gasoline.pcr$validation$PRESS / nrow(Xtr))
        1 comps 2 comps 3 comps 4 comps 5 comps 6 comps 7 comps
octane   1.5351  1.5802  1.6682 0.30099 0.30815 0.30309 0.26611
        8 comps 9 comps 10 comps
octane  0.27377 0.29495  0.28615
```

---

[1] Note that the y axis is labelled with "RMSEP" which sometimes is used in a more general sense—whether it deals with crossvalidation or true predictions must be deduced from the context.

The quality of the four-component model can be assessed by visualization: a very common plot shows the true values of the dependent variable on the x-axis and the predictions on the y-axis. We use a square plot (the right panel of Fig. 8.2) where both axes have the same range so that an ideal prediction would lie on a line with a slope of 45 degrees:

```
> par(pty = "s")
> plot(gasoline.pcr, "prediction", ncomp = 4)
> abline(0, 1, col = "gray")
```

Each point in the right plot of Fig. 8.2 is the prediction for that point when it was not part of the training set. Therefore, this plot gives some idea of what to expect for unseen data. Note, however, that this particular number of PCs was chosen with the explicit aim to minimize errors for these data points—the LOO crossvalidation was used to assess the optimal number of PCs. The corresponding error estimate is therefore optimistically biased, and we need another way of truly assessing the expected error for future observations.

This is given, for example, by the performance of the model on an unseen test set:

```
> gasoline.pcr.pred <- predict(gasoline.pcr, ncomp = 4,
+                              newdata = gasoline[gas.even, ])
> rms(gasoline$octane[gas.even], gasoline.pcr.pred)
[1] 0.21017
```

or, by using the RMSEP function again:

```
> RMSEP(gasoline.pcr, ncomp = 4, newdata = gasoline[gas.even, ],
+       intercept = FALSE)
[1]   0.2102
```

The error on the test set is 0.21, which is smaller than the cross validated error on the training set with four components (0.3010). Although this may seem surprising at first, it is again the result of the fact that LOO error estimates, although unbiased, have a large variance (Efron and Tibshirani 1993). We will come back to this point in Chap. 9.

Although the selection of the optimal number of components is a subjective process and should be guided by background knowledge as well as common sense, there are situations where a more automatic approach is desirable, e.g., in large-scale simulation studies. The **pls** package provides the function selectNcomp, implementing two different strategies. The first and simplest one estimates standard deviations for the RMSEP estimates in crossvalidation error curves such as shown in Fig. 8.2, and picks the simplest model that is within one standard deviation of the global optimum in the curve. This heuristical approach has been suggested in the context of methods like ridge regression and the lasso (Hastie et al. 2001) but can also be applied here. The second approach implements a permutation test (see also Sect. 9.3.1), again using the global minimum in the crossvalidation curve as a reference, and testing whether simpler models are significantly worse (Van der Voet 1994). Both approaches also provide graphical output. The input is a fitted model, including validation data:

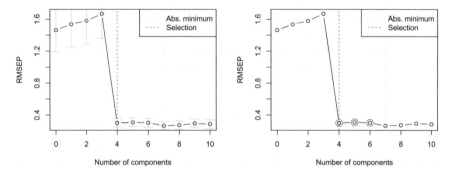

Fig. 8.3 Two strategies to automatically select the number of components in a PCR (or PLS) model from the **pls** package. Left plot: the one-sigma approach. The smallest model within one standard deviation from the optimal model (here at seven PCs) is picked. Right plot: the permutation test approach. Large blue circles indicate which models are compared to the global minimum. In this rather trivial example, both criteria choose four components

```
> nc.1s <-
+    selectNcomp(gasoline.pcr, method = "onesigma", plot = TRUE)
> nc.rand <-
+    selectNcomp(gasoline.pcr, method = "randomization", plot = TRUE)
```

For the gasoline data, the results are shown in Fig. 8.3—both methods select four components.

8.3   Partial Least Squares (PLS) Regression

In PCR, the information in the independent variables is summarized in a small number of principal components. However, there is no *a priori* reason why the PCs associated with the largest singular values should be most useful for regression. PC 1 covers the largest variance, but still may have only limited predictive power, as we have seen in the gasoline example. Since we routinely pick PCs starting from number 1 and going up, there is a real chance that we include variables that actually do not contribute to the regression model. Put differently: we compress information in X without regard to what is to be predicted, so we can never be sure that the essential part of the data is preserved. Although it has been claimed that *selecting* specific PCs (e.g., numbers 2, 5 and 6) on which to base the regression, rather than a sequence of PCs starting from one, leads to better models (see, e.g., Barros and Rutledge 1998), this only increases the difficulties one faces: selection is a much more difficult process than determining a threshold.

PLS forms an alternative. Just like PCR, PLS defines orthogonal latent variables to compress the information and throw away irrelevant stuff. However, PLS explicitly aims to construct latent variables in such a way as to capture most variance in X and Y, *and* to maximize the correlation between these matrices. Put differently: it maximizes the *covariance* between X and Y. So it seems we keep all the advantages, and get rid of the less desirable aspects of PCR. The algorithm is a bit more complicated than PCR; in fact, there exist several almost equivalent algorithms to perform PLS. The differences are caused by either small variations in the criterion that is optimized, different implementations to obtain speed improvements in specific situations, or by different choices for scaling intermediate results.

8.3.1   The Algorithm(s)

Just as in PCR, in PLS it is customary to perform mean-centering of the data so that there is no need to estimate an intercept vector; this is obtained afterwards. The notation is a bit more complicated than with PCR, as already mentioned, since now both X and Y matrices have scores and loadings. Moreover, in many algorithms one employs additional weight matrices. One other difference with PCR is that the components of PLS are extracted sequentially whereas the PCs in PCR can be obtained in one SVD step. In each iteration in the PLS algorithm, the variation associated with the estimated component is removed from the data in a process called deflation, and the remainder (indicated with E for the "deflated" X matrix, and F for the deflated Y) is used to estimate the next component. This continues until the user decides it has been enough, or until all components have been estimated.

The first component is obtained from an SVD of the crossproduct matrix $S = X^T Y$, thereby including information on both variation in X and Y, and on the correlation between both. The first left singular vector, w, can be seen as the direction of maximal variance in the crossproduct matrix, and is usually indicated with the somewhat vague description of "weights". The projections of matrix X on this vector are called "X scores":

$$t = Xw = Ew \tag{8.13}$$

Eventually, these scores t will be gathered in a matrix T that fulfills the same role as the score matrix in PCR; it is a low-dimensional, full-rank, estimate of the information in X. Therefore, regressing Y on T is easy, and the coefficient vector for T can be converted to a coefficient vector for the original variables.

The next step in the algorithm is to obtain loadings for X and Y by regressing against the *same* score vector t:

$$p = E^T t / (t^T t) \tag{8.14}$$

$$q = F^T t / (t^T t) \tag{8.15}$$

Notice that one divides by the sum of all squared elements in t: this leads to "normalized" loadings. It is not essential that the scaling is done in this way. In fact, there are numerous possibilities to scale either loadings, weights, or scores—one can choose to have either the scores or the loadings orthogonal. Unfortunately, this can make it difficult to compare the scores and loadings of different PLS implementations. The current description is analogous to PCR where the loadings are taken to have unit variance.

Finally, the data matrices are deflated: the information related to this latent variable, in the form of the outer products $t\,p^T$ and $t\,q^T$, is subtracted from the (current) data matrices.

$$E_{n+1} = E_n - t\,p^T \tag{8.16}$$

$$F_{n+1} = F_n - t\,q^T \tag{8.17}$$

The estimation of the next component then can start from the SVD of the crossproduct matrix $E_{n+1}^T F_{n+1}$. After every iteration, vectors w, t, p and q are saved as columns in matrices W, T, P and Q, respectively.

In words, the algorithm can be summarized as follows: the vectors w constitute the direction of most variation in the crossproduct matrix $X^T Y$. The scores t are the coordinates of the objects on this axis. Loadings for X and Y are obtained by regressing both matrices against the scores, and the products of the scores and loadings for X and Y are removed from data matrices E and F.

One complication is that columns of matrix W can not be compared directly: they are derived from successively deflated matrices E and F. An alternative way to represent the weights, in such a way that all columns relate to the original X matrix, is given by

$$R = W(P^T W)^{-1} \tag{8.18}$$

Matrix R has some interesting properties, one of which is that it is a generalized inverse for P^T. It also holds that $T = XR$. For interpretation purposes, one sometimes also calculates so-called y-scores $U = YQ$. Alternatively, these y-scores can be obtained as the right singular vectors of $E^T F$.

Now, we are in the same position as in the PCR case: instead of regressing Y on X, we use scores T to calculate the regression coefficients A, and later convert these back to the realm of the original variables:

$$Y = \tilde{X}B + \mathcal{E} = T(P^T B) + \mathcal{E} = TA + \mathcal{E} \tag{8.19}$$
$$A = (T^T T)^{-1} T^T Y \tag{8.20}$$
$$B = RA \tag{8.21}$$

These equations are almost identical with the PCR algorithm presented in Eqs. 8.7 and 8.8. The difference lies first and foremost in the calculation of T, which now includes information on Y, and in the calculation of the regression coefficients for the original variables, where PLS uses R rather than P. Again, the singularity problem is solved by using a low-dimensional score matrix T of full rank. The coefficient for the abscissa is obtained in the same way as with PCR (Eq. 8.11).

In the **pls** package, PLS regression is available as function `plsr`:

```
> gasoline.pls <- plsr(octane ~ ., data = gasoline,
+                     subset = gas.odd, ncomp = 5)
> summary(gasoline.pls)
Data:   X dimension: 30 401
Y dimension: 30 1
Fit method: kernelpls
Number of components considered: 5
TRAINING: % variance explained
          1 comps  2 comps  3 comps  4 comps  5 comps
X          71.71    79.70    90.71    95.70    96.59
octane     22.82    93.93    97.49    97.79    98.74
```

Clearly, the first components of the PLS model explain much more variation in Y than the corresponding PCR model: the first two PLS components already cover almost 94%, whereas the first two PCR components barely exceed ten percent. The price to be paid lies in the description of the X data: the two-component PCR model explains seven percent more than the corresponding PLS model.

To assess how many components are needed, the `validation` argument can be used, in the same way as with the `pcr` function:

```
> gasoline.pls <- plsr(octane ~ ., data = gasoline, subset = gas.odd,
+                     validation = "LOO", ncomp = 10)
> par(mfrow = c(1, 2))
> plot(gasoline.pls, "validation", estimate = "CV")
> par(pty = "s")
> plot(gasoline.pls, "prediction", ncomp = 3)
> abline(0, 1, col = "gray")
```

The resulting plots, shown in Fig. 8.4, indicate that a PLS model comparable to the four-component PCR model from Fig. 8.2 only needs three latent variables. This difference between PLS and PCR is often observed in practice: PLS models typically need one or two fewer components than PCR models to achieve similar CV error estimates.

Let's check the predictions of the unseen data, the even rows of the data frame:

```
> RMSEP(gasoline.pls, ncomp = 3, newdata = gasoline[gas.even, ],
+       intercept = FALSE)
[1]   0.2093
```

Indeed, with one component less the PLS model achieves a prediction quality that is as good as that of the PCR model.

The `plsr` function takes a `method` argument to specify which PLS algorithm is to be used. The default is `kernelpls` (Dayal and MacGregor 1997), a very fast and stable algorithm which gives results equal to the original NIPALS algorithm (Martens and Næs 1989) which is available as `oscorespls`. The kernel algorithm performs SVD on crossproduct matrix $X^T Y Y^T X$ rather than $X^T Y$, and

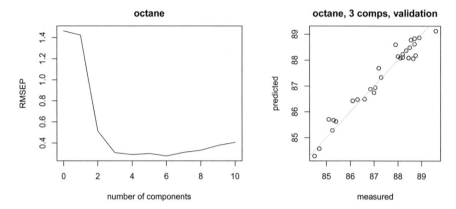

Fig. 8.4 Validation plot for PLS regression on the gasoline data (left) and prediction quality of the optimal model, containing three PLS components (right)

avoids deflation of Y. In cases with large numbers of variables (tens of thousands), a variant called `widekernelpls` (Rännar et al. 1994) is more appropriate—again, it operates by constructing a smaller kernel matrix, this time XX^TYY^T, on which to perform the SVD operations. However, it is numerically less stable than the default algorithm. Also the `widekernelpls` algorithm gives results that are identical (upon convergence) to the NIPALS results.

One popular alternative formulation, SIMPLS (de Jong 1993), deflates matrix S rather than matrices E and F individually. It can be shown that SIMPLS actually maximizes the covariance between X and Y (which is usually taken as "the" PLS criterion), whereas the other algorithms are good approximations; for univariate Y, SIMPLS predictions are equal to the results from NIPALS and kernel algorithms. However, for multivariate Y, there may be (minor) differences between the approaches. SIMPLS can be invoked by providing the `method = "simpls"` argument to the `plsr` function. In all these variants, the scores will be orthogonal, whereas the loadings are not:

```
> cor(gasoline.pls$loadings[, 1:3])
          Comp 1     Comp 2     Comp 3
Comp 1  1.000000 -0.553292 -0.075284
Comp 2 -0.553292  1.000000 -0.062259
Comp 3 -0.075284 -0.062259  1.000000


> cor(gasoline.pls$scores[, 1:3])
           Comp 1      Comp 2      Comp 3
Comp 1  1.0000e+00 -1.4330e-16 -9.5352e-17
Comp 2 -1.4330e-16  1.0000e+00  6.4193e-17
Comp 3 -9.5352e-17  6.4193e-17  1.0000e+00
```

Given that one has a certain freedom to decide where exactly in the algorithm to normalize, the outcome of different implementations, and in particular, in different software packages, may seem to vary significantly. However, the regression coefficients, and therefore the predictions, of all these are usually virtually identical. For all practical purposes there is no reason to prefer the outcome of one algorithm over another.

8.3.2 Interpretation

PLS models give separate scores and loadings for X and Y, and additionally, in most implementations, some form of a weight matrix. In most cases, one concentrates on the matrix of regression coefficients B which is independent of algorithmic details such as the exact way of normalization of weights, scores and loadings, and is directly comparable to regression coefficients from other methods like PCR. Sometimes, plots of weights, loadings or scores of individual components can be informative, too, although one should be careful not to overinterpret: there is no a priori reason to assume that the individual components directly correspond to chemically interpretable entities (Frank and Friedman 1993).

The interpretation of the scores and loadings is similar to PCA: a score indicates how much a particular object contributes to a latent variable, while a loading indicates the contribution of a particular *variable*. An example of a loading plot is obtained using the code below:

```
> plot(gasoline.pls, "loading", comps = 1:3, legendpos = "topleft",
+     lty = 1, col = 1:3)
```

This leads to Fig. 8.5. The percentage shown in the legend corresponds with the variation explained of the X matrix for each latent variable. Note that the third component explains more variation of X than the second; in a PCR model this would be impossible.[2] Components one and two show spectrum-like shapes, with the largest values at the locations of the main features in the data, as expected—the third component is focusing very much on the last ten data points. This raises questions on the validity of the model: it is doubtful that these few (and noisy) wavelengths should play a major part. Perhaps a more prudent choice for the number of latent variables from Fig. 8.4 would have been to use only two.

Biplots, showing the relations between scores and loadings, can be made using the function `biplot.mvr`. One can in this way inspect the influence of individual PLS components, where the regression coefficient matrix B gives a more global view summarizing the influence of all PLS components. The argument `which`, taking the values `"x"`, `"y"`, `"scores"` and `"loadings"`, indicates what type of biplot is required. In the first case, the scores and loadings for x are shown in a biplot; the

---

[2]The `plot.mvr` function can be applied to PCR models as well as PLS models, so the discussion in this paragraph pertains to both.

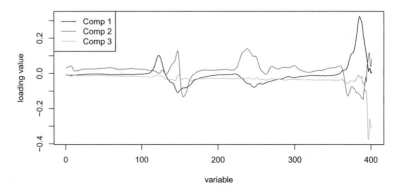

Fig. 8.5 PLS loadings for the first three latent variables for the gasoline data; the third component has large loadings only for the last ten variables

second case does the same for y. The other two options show combinations of x- and y- scores and loadings, respectively. In the current example, y-loadings are not very interesting since there is only one y variable, octane. An additional source of information is the relation between the X-scores, T, and the Y-scores, U. For the gasoline model these plots are shown for the first two latent variables in Fig. 8.6 using the following code:

```
> plot(scores(gasoline.pls)[, 1], Yscores(gasoline.pls)[, 1],
+      xlab = "X scores", ylab = "Y scores", main = "LV 1")
> abline(h = 0, v = 0, col = "gray")
> plot(scores(gasoline.pls)[, 2], Yscores(gasoline.pls)[, 2],
+      xlab = "X scores", ylab = "Y scores", main = "LV 2")
> abline(h = 0, v = 0, col = "gray")
```

Usually one hopes to see a linear relation, as is the case for the second latent variable; the first LV shows a less linear behavior. One could replace the linear regression in Eqs. 8.14 and 8.15 by a polynomial regression (using columns of powers of t), or even a non-linear regression. There are, however, not many reports where this has led to significant improvements (see, e.g., Wold et al. 1989; Hasegawa et al. 1996).

For multivariate Y, there is an additional difference between PLS and PCR. With PCR, separate models are fit for each Y variable: the algorithm does not try to make use of any correlation between the separate dependent variables. With PLS, this is different. Of course, one can fit separate PLS models for each Y variable (this is often indicated with the acronym PLS1), but one can also do it all in one go (PLS2). In that case, the same X-scores T are used for *all* dependent variables; although a separate set of regression coefficients will be generated for every y-variable, implicitly information from the other dependent variables is taken into account. This can be an advantage, especially when there is appreciable correlation between the y-variables, but in practice there is often little difference between the two approaches. Most people prefer multiple PLS1 models (analogous to PCR regression) since they seem to give slightly better fits.

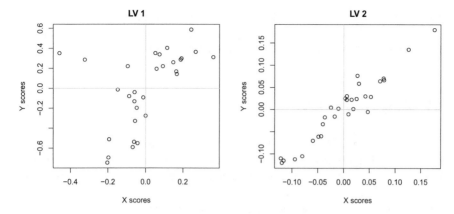

Fig. 8.6 Relation between X-scores and Y-scores, T and U, respectively, for the first two latent variables (gasoline data)

It is no coincidence that chemistry has been the first area in which PLS was really heavily used. Analytical chemists had been measuring spectra and trying to relate them to chemical properties for years, when this regression technique finally provided them with the tool to do so. Other disciplines such as statistics were slow to follow, but eventually PLS has found its place in a wide range of fields. Also the theoretical background has now been clarified: PLS started out as an algorithm that was really poorly understood. Nowadays, there are very few people who dispute that PLS is an extremely useful method, but it has been overhyped somewhat. In practical applications, its performance is very similar to techniques like PCR. Careful thinking of experimental design, perhaps variable selection, and appropriate preprocessing of the data is likely to be far more important than the exact choice of multivariate regression technique.

8.3.2.1 PLS Packages for R

Apart from the **pls** package, several other packages provide PLS functions, both for regression and classification. The list is large and rapidly expanding; examples include the packages **lspls** implementing Least-Squares PLS (Jorgensen et al. 2004), and **gpls** (Ding and Gentleman 2005) implementing generalized partial least squares, based on the Iteratively ReWeighted Least Squares (IRWLS) method (Marx 1996). Weighted-average PLS (ter Braak and Juggins 1993), often used in paleolimnology, can be found in package **paltran**. Package **plsgenomics** implements methods for classification with microarray data and prediction of transcription factor activities combining, amongst others, ridge regression with PLS. This package also provides function pls-lda performing LDA on PLS scores (see Sect. 8.7.3). Package **plspm** contains, in addition to the usual functions for PLS regression, also functions for Path Modelling (Tenenhaus et al. 2005). Penalized PLS (Krämer et al. 2008) is available in package **ppls**, and sparse PLS, forcing small loadings to become zero so that fewer variables are taking part in the model, in package **spls** (Chun and Keles 2010). Which of these packages is most suited depends on the application.

8.4 Ridge Regression

Ridge Regression (RR, Hoerl 1962; Hoerl and Kennard 1970) is another way to tackle regression problems with singular covariance matrices, usually for univariate Y. From all possible regression models giving identical predictions for the data at hand, RR selects the one with the smallest coefficients. This loss function is implemented by posing a (quadratic) penalty on the size of the coefficients B:

$$\underset{B}{\operatorname{argmax}} \ (Y - XB)^2 + \lambda B^T B \tag{8.22}$$

The solution is given by

$$\hat{B} = \left(X^T X + \lambda I\right)^{-1} X^T Y \tag{8.23}$$

Compared to Eq. 8.2, a constant is added to the diagonal of the crossproduct matrix $X^T X$, which makes it non-singular. The size of λ is something that has to be determined (see below). This *shrinkage* property has obvious advantages in the case of collinearities: even if the RR model is not quite correct, it will not lead to wildly inaccurate predictions (which may happen when some coefficients are very large). Usually, the intercept is not included in the penalization: one would expect that adding a constant c to the y-values would lead to predictions that are exactly the same amount larger. If the intercept would be penalized as well, this would not be the case.

Optimal values for λ may be determined by crossvalidation (or variants thereof). Several other, direct, estimates of optimal values for λ have been proposed. Hoerl and Kennard (Hoerl and Kennard 1970) use the ratio of the residual variance s^2, estimated from the model, and the largest regression coefficient:

$$\hat{\lambda}_{HK} = \frac{s^2}{\max(B_i^2)} \tag{8.24}$$

A better estimate is formed by using the harmonic means of the regression coefficients rather than the largest value. This is known as the Hoerl-Kennard-Baldwin estimate (Hoerl et al. 1975):

$$\hat{\lambda}_{HKB} = \frac{ps^2}{B^T B} \tag{8.25}$$

where p, as usual, indicates the number of columns in X. A variance-weighted version of the latter is given by the Lawless-Wang estimate (Lawless and Wang 1976):

$$\hat{\lambda}_{LW} = \frac{ps^2}{B^T X^T X B} \tag{8.26}$$

Since PCR and PLS can also be viewed as shrinkage methods (Frank and Friedman 1993; Hastie et al. 2001), there are interesting links with ridge regression. All three shrink the regression coefficients away from directions of low variation. Ridge regression can be shown to be equivalent to PCR with shrunken eigenvalues for the principal components; PCR uses a hard threshold to select which PCs to take into account. PLS also shrinks—it usually takes the middle ground between PCR and RR. However, in some cases PLS coefficients may be inflated, which may lead to slightly worse performance (Hastie et al. 2001). In practice, all three methods lead to very similar results. Another common feature is that just like PCR and PLS, ridge regression is not affine equivariant: it is sensitive to different (linear) scalings of the input. In many cases, autoscaling is applied by default. It is also hard-coded in the lm.ridge function in package **MASS**. Unfortunately, for many types of spectroscopic data autoscaling is not very appropriate, as we have seen earlier. For the gasoline data, it does not work very well either:

```
> gasoline.ridge <-
+    lm.ridge(octane ~ NIR, data = gasoline, subset = gas.odd,
+              lambda = seq(0.001, 0.1, by = 0.01))
> select(gasoline.ridge)
modified HKB estimator is -6.256e-28
modified L-W estimator is -3.6741e-28
smallest value of GCV at 0.001
```

Both the HKB estimate and the L-W estimate suggest a very small value of λ; the generalized crossvalidation (see Chap. 9) suggests the smallest value of λ is the best.

Apart from the links with PCR and PLS, ridge regression is also closely related to SVMs when seen in the context of regression—we will come back to this in Sect. 8.6.1. Related methods using L_1 penalization, such as the lasso and the elastic net, will be treated in more detail in Sect. 10.2.

8.5   Continuum Methods

In many ways, MLR and PCR form the opposite ends of a scale. In PCR, the stress is on summarizing the variance in X; the correlation with the property that is to be predicted is not taken into account in defining the latent variables. With MLR the opposite is true: one does not care how much information in X is actually used as long as the predictions are OK. PLS takes a middle ground with the criterion that latent variables should explain as much of the covariance between X and Y as possible.

There have been attempts to create regression methods that offer other intermediate positions, most notably Continuum Regression (CR, Stone and Brooks 1990) and Principal Covariates Regression (PCovR, de Jong and Kiers 1992). Although they are interesting from a theoretical viewpoint, in practice they have never caught on.

One possible explanation is that there is little gain in yet another form of multivariate regression, where methods like PCR, PLS and RR already exist and in most cases give very similar results. Moreover, these continuum methods provide additional crossvalidation problems because more often than not an extra parameter needs to be set.

8.6 Some Non-linear Regression Techniques

Many non-linear techniques are available for regression. Here, we will very briefly focus on two classes of methods that we have already seen in Chap. 7 on classification, SVMs and neural networks. Both can be adapted to continuous output without much trouble.

8.6.1 SVMs for Regression

In order not to get lost in mathematical details that would be out of context in this book, only the rough contours of the use of SVMs in regression problems are sketched here. A more thorough treatment can be found in the literature (Hastie et al. 2001). Typically, SVMs tackle linear regression problems by minimization of a loss function of the following form:

$$L_{\text{SVM}} = \sum_i V(y_i - f(x_i)) + \lambda ||\beta||^2 \tag{8.27}$$

where the term $V(y_i - f(x_i))$ corresponds to an error function describing the differences between experimental and fitted values, and the second term is a regularization term, keeping the size of the coefficients small. If the error function V is taken to be the usual squared error then this formulation is equal to ridge regression, but more often other forms are used. Two typical examples are shown in Fig. 8.7: the left panel shows a so-called ε-insensitive error function, where only errors larger than a cut-off ε are taken into account (linearly), and the right plot shows the Huber loss function, which is quadratic up to a certain value c and linear above that value.

The reason to use these functions is that the linear error function for larger errors leads to more robust behavior. Moreover, the solution of the loss function in Eq. 8.27 can be formulated in terms of inner products, just like in the classification case, and again only a subset of all coefficients (the support vectors) are non-zero. Moreover, kernels can be used to find simple linear relationships in high dimensions which after back-transformation represent much more complex patterns.

Let us concentrate on how to use the svm function, seen earlier, in regression problems. We again take the gasoline data, and fit a model, using ten-fold crossvalidation:

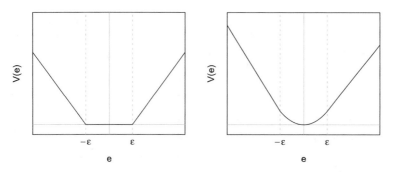

Fig. 8.7 Typical error functions for SVMs in a regression setting: left, the ε-insensitive error function; right, the Huber function

```
> gasoline.svm <- svm(octane ~ ., data = gasoline,
+                     subset = gas.odd, cross = 10)
> (gasoline.svmsum <- summary(gasoline.svm))

Call:
svm(formula = octane ~ ., data = gasoline, cross = 10,
    subset = gas.odd)


Parameters:
   SVM-Type:  eps-regression
 SVM-Kernel:  radial
       cost:  1
      gamma:  0.0024938
    epsilon:  0.1


Number of Support Vectors:  27


10-fold cross-validation on training data:

Total Mean Squared Error: 0.54107
Squared Correlation Coefficient: 0.79318
Mean Squared Errors:
 0.82181 0.046941 1.0881 0.48363 0.10566 1.0402 0.098511
 0.43532 0.29603 0.99446
```

Clearly, the crossvalidation leads to a relatively large error, corresponding to an RMSCV of 0.7356. Note that the `svm` function by default performs autoscaling on both X and y, which for the gasoline data is not optimal for reasons discussed earlier. Plotting the predictions for the training and test sets clearly shows the difference in quality:

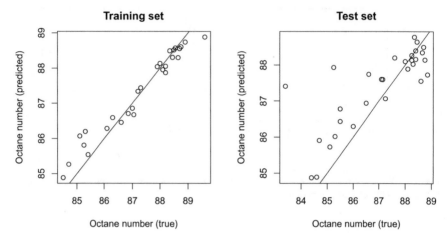

Fig. 8.8 Predictions for the training data (left) and the test data (right) using the default values of the svm function

```
> plot(gasoline$octane[gas.odd], predict(gasoline.svm),
+       main = "Training set", xlab = "Octane number (true)",
+       ylab = "Octane number (predicted)")
> abline(0, 1)
> plot(gasoline$octane[gas.even],
+       predict(gasoline.svm, new = gasoline[gas.even, ]),
+       main = "Test set", xlab = "Octane number (true)",
+       ylab = "Octane number (predicted)")
> abline(0, 1)
```

The result, shown in Fig. 8.8, tells us that the training data are predicted quite well (although there seems to be a slight tendency to predict too close to the mean), but the predictions for the test data show large errors. The corresponding RMS value for prediction is:

```
> rms(gasoline$octane[gas.even],
+       predict(gasoline.svm, new = gasoline[gas.even, ]))
[1] 1.0671
```

Apparently, the model is not able to generalize, which should be no surprise given the already quite large crossvalidation error seen in the summary, both clear indications that the model is overfitting.

In this kind of situation, one should consider the parameters of the method. The default behavior of the svm function is to use the ε-insensitive error function, with $\varepsilon = .1$, a value of 1 for the penalization factor in Eq. 8.27, and a gaussian ("radial basis") kernel, which also has some parameters to tune. Since in this case already many variables are available for the SVM, it makes sense to use a less flexible kernel. Indeed, swapping the radial basis kernel for a linear one makes a big difference for the prediction of the test data:

```
> gasoline.svmlin <- svm(octane ~ ., data = gasoline,
+                        subset = gas.odd, kernel = "linear")
> rms(gasoline$octane[gas.even],
+     predict(gasoline.svmlin, new = gasoline[gas.even, ]))
[1] 0.26418
```

Optimization of the parameters, for instance using `tune.svm`, should lead to further improvements.

8.6.2 ANNs for Regression

The basic structure of a backpropagation network as shown in Fig. 7.13 remains unchanged for regression applications: the numbers of input and output units equal the number of independent and dependent variables, respectively. Again, the number of hidden units is subject to optimization. In high-dimensional cases one practical difficulty needs to be solved—the data need to be compressed, otherwise the number of weights would be too high. In many practical applications, PCA is performed on the input matrix and the network is trained on the scores.

Let's see how that works out for the gasoline data. We will use the scores of the first five PCs, and fit a neural network model with five hidden units. To indicate that we want numerical output rather than class output, we set the argument `linout` to TRUE (the default is use logistic output units):

```
> X <- scale(gasoline$NIR, scale = FALSE,
+            center = colMeans(gasoline$NIR[gas.odd, ]))
> X.odd.svd <- svd(X[gas.odd, ])
> X.odd.scores <- X.odd.svd$u %*% diag(X.odd.svd$d)
> X.even.scores <- X[gas.even, ] %*% X.odd.svd$v
> gas.nnet <- nnet(X.odd.scores[, 1:5],
+                  matrix(gasoline$octane[gas.odd], ncol = 1),
+                  size = 5, linout = TRUE)
# weights:  36
initial  value 228545.146358
iter  10 value 52.125172
iter  20 value 25.813681
iter  30 value 7.492666
iter  40 value 2.621245
iter  50 value 1.899539
iter  60 value 1.715577
iter  70 value 1.647011
iter  80 value 1.636999
iter  90 value 1.619624
iter 100 value 1.613594
final  value 1.613594
stopped after 100 iterations
```

The number of weights is 36, perhaps already a bit too high given the limited number of samples. Clearly, the decrease in error value for the training set has not yet slowed

Fig. 8.9 Histogram of
prediction errors (RMS
values) for the even rows of
the gasoline data upon
repeated neural network
training

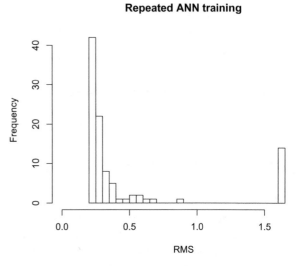

after the (default) maximum value of 100 training iterations—however, using more
iterations may lead to overfitting. The usual approach is to divide the data in three
sets: the training set, the validation set, of which the predictions are continuously
monitored, and a test set. As soon as the errors in the predictions for the validation
set start to increase, training is stopped. At that point, the network is considered to
be trained, and its prediction errors can be assessed using the test set. Obviously,
this can only be done when the number of data points is not too small, a serious
impediment in the application of neural nets in the life sciences.

For the moment, we will ignore the issue, and we will assess the predictive performance of our final network:

```
> gas.nnet.pred <- predict(gas.nnet, X.even.scores)
> rms(gas.nnet.pred, gasoline$octane[gas.even])
[1] 0.21797
```

The prediction error is similar to the PCR result in Sect. 8.2.2. Obviously, not only
the number of training iterations, but also the number of hidden units needs to
be optimized. Similar to the approach in classification, the convenience function
tune.nnet can be helpful.

One further remark needs to be made: since the initialization of neural nets usually is done randomly, repeated training sessions rarely lead to comparable results.
Figure 8.9 shows a histogram of RMS prediction errors (test set, even rows of the
gasoline data) of 100 trained networks. Clearly, there is considerable spread. Even a
local optimum can be discerned around 1.63 in which more than 10% of the networks
end up. The bottom line is that although neural networks have great modelling power,
one has to be very careful in using them—in particular, one should have enough data
points to allow for rigid validation.

8.7 Classification as a Regression Problem

In many cases, classification can be tackled as a regression problem. The obvious example is logistic regression, one of the most often used classification methods in the medical and social sciences. In logistic regression, one models the log of the *odds ratio*, usually with a multiple linear regression:

$$\ln \frac{p}{1-p} = XB \tag{8.28}$$

where p is the probability of belonging to class 1. This way of modelling has several advantages, the most important of which may be that the result can be directly interpreted as a probability—the p value will always be between 0 and 1. Moreover, the technique makes very few assumptions: the independent variables do not need to have constant variance, or even be normally distributed; they may even be categorical. The usual least-squares estimators for finding B are not used, but rather numerical optimization techniques maximizing the likelihood. However, a big disadvantage is that many data points are needed. Because of the high dimensionality of most data sets in the life sciences, logistic regression has not been widely adopted in this field and we will not treat it further here.

Although it may not seem immediately very useful to use regression for classification problems, it does open up a whole new field of elaborate, possibly non-linear regression techniques as classifiers. Another very important application is classification of "fat" data matrices, or data sets with many more variables than objects. These have become the rule rather than the exception in the natural sciences. Although it means that a lot of information is available for each sample, it also means in practice that a lot of numbers are available that do not say anything particularly interesting about the sample—these can be pure noise, but also genuine signals, unrelated to the research question at hand. Another problem is the correlation that is often present between variables. Finding relevant differences between classes of samples in such a situation is difficult: the number of parameters to estimate in regular forms of discriminant analysis far exceeds the number of independent samples available. An example is the `prostate` data set, containing more than 10,000 variables and only 327 samples. We could eliminate several without losing information, and also removing variables that are "clearly" not related to the dependent variable (in as far as we would be able to recognize these) would help, the idea that is formalized in the OPLS approach discussed in Sect. 11.4. An alternative is formed by variable selection techniques, such as the ones described in Chap. 10, but these usually rely on accurate error estimates that are hard to get with low numbers of samples.

The same low number of samples also forces the statistical models to have very few parameters: fat matrices can be seen as describing sparsely—very sparsely—populated high-dimensional spaces, and only the simplest possible models have any chance of gaining predictive power. The simplest possible case is that of linear discriminant analysis, but direct calculation of the coefficients is impossible because of the matrix inversion in Eq. 7.6—the covariance matrix is singular. Regularization approaches like RDA are one solution; the extreme form of regularization, diagonal LDA, enjoys great popularity in the realm of microarray analysis. Another often-used strategy is to compress the information in a much smaller number of variables, usually linear combinations of the original set, and perform simple methods like LDA on the new, small, data matrix. Two approaches are popular: PCDA and PLSDA. We start, however, by showing the general idea of using regression for classification purposes in the context of LDA.

8.7.1   Regression for LDA

In the case of a two-class problem (healthy/diseased, true/false, yes/no) the dependent variable is coded as 1 for the first class, and 0 for the other class. For prediction, any object whose predicted value is above 0.5 will be classified in the second class. In the field of machine learning, often a representation is chosen where one class is indicated with the label -1 and the other one with 1; the class boundary then is at 0. For problems involving more than two classes, a class matrix is used with one column for every class and at most one "1" per row; the position of the "1" indicates the class of that particular object. We saw this already in the chapter on classification. In principle, one could even remove one of these columns: since every row adds to one (every object is part of a class) the number of independent columns is the number of classes minus one. In practice, however, this is rarely done.

   To illustrate the close connection between discriminant analysis and linear regression, consider a two-variable subset of the wine data with equal sizes of only classes Barolo and Grignolino:

```
> C <- classvec2classmat(vintages[c(1:25, 61:85)])
> X <- wines[c(1:25, 61:85), c(7, 13)]
```

The regression model can be written as

$$C = XB + \mathcal{E} \tag{8.29}$$

where C is the two-column class matrix—note that because this is only a two-class problem, we could also have used one vector with the 0/1 or $-1/1$ coding. Solving this equation by least squares leads to:

```
> C <- classvec2classmat(vintages[c(1:25, 61:85)])
> X <- wines[c(1:25, 61:85), c(7, 13)]
> wines.lm <- lm(C ~ X)
> wines.lm.predict <- classmat2classvec(predict(wines.lm))
> table(vintages[c(1:25, 61:85)], wines.lm.predict)
            wines.lm.predict
            Barbera Barolo Grignolino
  Barbera         0      0          0
  Barolo          0     23          2
  Grignolino      0      1         24
```

This is exactly the same classification as the one obtained from LDA:

```
> wines.lda <- lda(factor(vintages[c(1:25, 61:85)]) ~ X)
> table(vintages[c(1:25, 61:85)], predict(wines.lda)$class)

            Barolo Grignolino
  Barbera        0          0
  Barolo        23          2
  Grignolino     1         24
```

The factor function in the lda line is used to convert the three-level factor vintages to a two-level factor containing only Barolo and Grignolino, making the comparison with the lm predictions easier. This means that instead of doing LDA, we could use linear regression with a binary dependent variable: any object for which the predicted value is larger than 0.5 will be classified in class 2, otherwise in class 1. The direct equality of the least-squares solution with LDA only holds for two groups with equal class sizes (Ripley 1996; Hastie et al. 2001); for more classes, or classes with different sizes, the linear regression approach actually optimizes a slightly different criterion than LDA.

8.7.2 PCDA

One way to compress the information in a fat data matrix into something that is more easy to analyse is PCA. Subsequently, LDA is performed on the scores—the result is often referred to as PCDA or PCLDA. The prostate data provide a nice example: the number of variables far exceeds the number of samples, even though that number is not low in absolute terms. Again, we are trying to discriminate between control samples and cancer samples, so we consider only two of the three classes. Now, however, we use all variables—in the cases of SVMs and boosting this would have led to large memory demands, but the current procedure is much more efficient. Using the SVD on the crossproduct matrix of the non-bph samples of the prostate data, similar to the procedure shown in Sect. 4.2, we obtain scores and loadings. The first sixteen PCs cover just over 70% of the variance of the X matrix, not too surprising given the number of variables. We should have a look at the scores, for clarity limiting ourselves to the first five components:

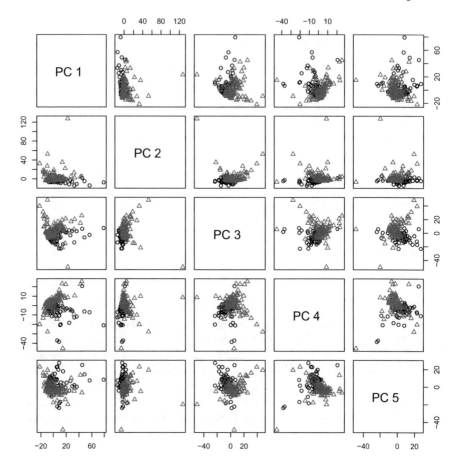

Fig. 8.10 Pairs plot of the scores of the prostate data in the first five PCs. Red triangles indicate cancer samples, black circles are controls

```
> pairs(prost.scores[, 1:5], labels = paste("PC", 1:5),
+       pch = as.integer(prost.type), col = as.integer(prost.type))
```

The result is shown in Fig. 8.10. Although some interesting structure is visible, there is no obvious separation between the classes in any of the plots. This five-dimensional representation of the data can be used in any form of discriminant analysis; we will stick to LDA, and just to get a feeling for what we can hope to expect, we will use five PCs. The naive, and as we shall see later, incorrect approach would be the following:

```
> ## INCORRECT
> prost.pcda5 <- lda(prost.type ~ prost.scores[, 1:5], CV = TRUE)
> (prost.ldaresult <- table(prost.type, prost.pcda5$class))

prost.type control pca
   control      51   30
      pca        8  160
```

Leave-one-out crossvalidation leads to a correct prediction in 85% of the cases, better than we might have expected on the basis of Fig. 8.10. One should not forget, however, that the cancer class is more than twice the size of the control class, so that already a not-too-clever random classification of cancer for all samples would lead to a success rate of over 65%. Note that the prediction errors are slightly unbalanced: more control samples are predicted to be cancer than vice versa. This is the result of the default prior of the lda function, which is proportional to the class representation in the training set.

As already stated, the above procedure is incorrect: the error estimate is optimistically biased because the PCA step (including mean-centering and scaling) has not been incorporated in the crossvalidation. As it is now, the left-out sample still exerts influence on the classification model through its contribution to the PCs, whereas in the correct way, the crossvalidation should include the PCA. This can be done by using an explicit crossvalidation loop, leaving out part of the samples, performing PCA and building the LDA model, but a more easy approach is to see the classification as a regression problem and use the pcr function, with its built-in crossvalidation facilities. While we are at it, we should also separate training data from test data, in order to get some kind of estimate for the prediction error, as well as the optimal number of latent variables. After rearranging the data into a form that fits the pcr function, application is simple:

```
> prost.df2 <- data.frame(class = as.integer(prost.type),
+                         msdata = I(prost))
> prost.pcr <- pcr(class ~ msdata, ncomp = 16,
+                   data = prost.df2, subset = prost.odd,
+                   validation = "CV", scale = TRUE)
> validationplot(prost.pcr, estimate = "CV")
```

In this case the validation (by default a ten-fold crossvalidation) is done correctly: the scaling and the PCA are done only *after* the out-of-bag samples are removed from the training data. This leads to the validation plot in the left panel of Fig. 8.11—a cautious person would probably select five PCs here, but since the number of samples is quite large, one might even consider eleven PCs.

Note that the RMSEP measure shown here is not quite what we are interested in: rather than the deviation from the ideal values of 0 and 1 in the classification matrix, we should look at the number of correct classifications (Kjeldahl and Bro 2010). For the training data this is achieved by directly assessing the cross-validated predictions in the mvr object returned by the pcr function:

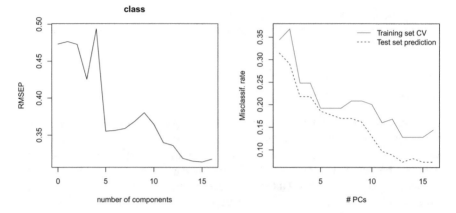

Fig. 8.11 Validation plots for the PCR regression plot of the prostate MS data: discrimination between `control` and `pca` samples. Left plot: RMSEP values for the class codes. Right plot: fraction of misclassifications, for the cross-validated training set predictions, and prediction of new, unseen data

```
> prost.cv.cl <- round(prost.pcr$validation$pred[, 1, ])
> prost.cv.err <- apply(prost.cv.cl, 2,
+                       err.rate, prost.df2$class[prost.odd])
```

Predictions for the test data can be achieved by the by now well-known `predict` function:

```
> prost.tst <- predict(prost.pcr, newdata = prost.df2[prost.even, ])
> prost.tst.cl <- round(prost.tst[, 1, ])
> prost.tst.err <- apply(prost.tst.cl, 2,
+                        err.rate, prost.df2$class[prost.even])
>
> matplot(cbind(prost.cv.err, prost.tst.err),
+         lty = 1:2, col = 2:1, type = "l",
+         xlab = "# PCs", ylab = "Misclassif. rate")
> legend("topright", lty = 1:2, col = 2:1, bty = "n",
+        legend = c("Training set CV", "Test set prediction"))
```

The result is shown in the right plot of Fig. 8.11. The two lines clearly follow a very similar trajectory, confirming that crossvalidation is a quite reliable way of estimating errors, or in this case misclassification rates: there is no evidence here of overfitting. Normal choices for the optimal number of PCs again (looking only at the training set CV line!) would be five (the first local minimum), eleven, or even thirteen. The test data show that with eleven PCs the expected classification error is around 10%.

One big advantage of the regression approach to classification is that it is not restricted to two-class problems only, but can be directly applied to multiclass problems. Rather than presenting a factor as the dependent variable, we use the matrix-membership notation and present a matrix, with one column per class. For the PCR-based approach, application to the full three-class prostate data (only 1,000 variables, however) is achieved as follows:

```
> prostate.clmat <- classvec2classmat(prostate.type)
> prostate.df <- data.frame(class = I(prostate.clmat),
+                           msdata = I(prostate[, 1:1000]))
> prostate.odd <- seq(1, nrow(prostate.df), by = 2)
> prostate.even <- seq(2, nrow(prostate.df), by = 2)
> prostate.pcr <- pcr(class ~ msdata, ncomp = 16,
+                     data = prostate.df,
+                     subset = prostate.odd,
+                     validation = "CV", scale = TRUE)
```

Again, we should convert the predicted values in the PCR crossvalidation to classes, and plot the number of misclassifications so that we can pick the optimal number of components:

```
> pcr.predictions.loo <-
+    sapply(1:16,
+           function(i, arr) classmat2classvec(arr[, , i]),
+           prostate.pcr$validation$pred)
> pcr.loo.err <- apply(pcr.predictions.loo, 2, err.rate,
+                      prostate.type[prostate.odd])
```

Test set results can be obtained in a completely analogous way:

```
> prostate.pcrpred <-
+    predict(prostate.pcr, new = prostate.df[prost.even, ])
> predictions.pcrtest <-
+    sapply(1:16,
+           function(i, arr) classmat2classvec(arr[, , i]),
+           prostate.pcrpred)


> matplot(cbind(pcr.loo.err,
+               apply(predictions.pcrtest, 2, err.rate,
+                     prostate.type[prost.even])),
+         type = "l", lty = 1:2, col = 2:1,
+         main = "PCDA",
+         xlab = "# PCs", ylab = "Misclassif. rate")
> legend("topright", lty = 1:2, col = 2:1, bty = "n",
+        legend = c("Training set CV", "Test set prediction"))
```

The left panel in Fig. 8.12 shows the result. The solid red line corresponds to the CV estimates of misclassification rate; the black dashed line to the results of the test set predictions. Clearly, the classification error is quite high for all numbers of components. We could ask ourselves what is going wrong, and focus on, e.g., seven components:

```
> table(prostate.type[prost.even], predictions.pcrtest[, 7])
```

```
         bph control pca
bph        1       0  38
control    0      30  10
pca        0       4  41
```

There is considerably confusion between the pca and bph classes: almost all bph objects are classified as pca. The controls are separated relatively well.

8.7.3 PLSDA

Although the above approach often is reported to work well in practice, it has a (familiar) flaw: there is no reason to assume that the information relevant for the class discrimination is captured in the *first* PCs. Since PLS takes into account the dependent variable when defining latent variables, this is a logical alternative. In literature, this form of discriminant analysis, usually done in the form of a direct regression on coded class variables, is called PLSDA (Barker and Rayens 2003). For the prostate data, this leads to:

```
> prostate.pls <- plsr(class ~ msdata, data = prostate.df,
+                      subset = prostate.odd, scale = TRUE,
+                      ncomp = 16, validation = "CV")
```

Using code that is completely analogous to the PCDA case on the previous pages, we arrive at the right plot in Fig. 8.12 (note the difference in scales at the y-axis!). As expected, fewer components are needed for optimal results—four would be selected in the case of PLSDA, whereas PCR would need seven. The PLSDA error values are considerably lower than with PCR, both for the LOO crossvalidation and for the test data: already with one PLS component the prediction of the test data is better than the PCDA model achieves with eight. With four components, PLSDA prediction of the test set looks like this:

```
> table(prostate.type[prostate.even], predictions.plstest[, 4])
```

```
         bph control pca
bph       10       0  29
control    0      35   5
pca        4       2  78
```

Clearly, the confusion between the bph and pca classes is still there, but results are somewhat better.

An alternative is to perform a classical LDA on the PLS scores—in several papers (e.g., Barker and Rayens 2003; Nguyen and Rocke 2002; Boulesteix 2004) this is claimed to be superior in quality. Where performing LDA on the first couple of PCs is completely analogous to doing PCR on the class matrix (for equal class sizes, at least),

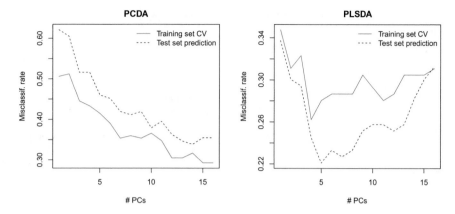

Fig. 8.12 PCDA (left) and PLSDA (right) classification results for the three-class prostate data

this is not the case in PLS, since class knowledge is used in defining latent variables. One therefore may expect some differences. First, we build an LDA model on the scores of the training set, directly available with the `scores` extractor function:

```
> prostate.ldapls <- lda(scores(prostate.pls)[, 1:6],
+                        prostate.type[prostate.odd])
```

Next, we can build an LDA model on the scores of the training set, directly available with the `scores` extractor function, and use this model to make predictions for the test set:

```
> tst.scores <- predict(prostate.pls, ncomp = 1:6,
+                       newdata = prostate.df[prostate.even, ],
+                       type = "scores")
> table(prostate.type[prostate.even],
+       predict(prostate.ldapls, new = tst.scores)$class)

          bph control pca
  bph      24       0  15
  control   3      33   4
  pca      15       1  68
```

Compared to the direct PLSDA method, the total number of misclassifications is the same, although slightly different for individual classes. For this case at least, the differences are small.

8.7.3.1 A Word of Warning

Because it is more focused on information in the dependent variable, PLS can be called a more greedy algorithm than PCR. In many cases this leads to a better fit for PLS (with the same number of components as the PCR model, that is), but it also presents a bigger risk of overfitting. The following example will make this clear: suppose we generate random data from a normal distribution, and allocate every sample randomly to one of two possible classes:

```
> nvar <- 2000
> nobj <- 40
> RandX <- matrix(rnorm(nobj*nvar), nrow = nobj)
> RandY <- sample(c(0, 1), nobj, replace = TRUE)
```

Next, we compress the information in variable RandX into two latent variables, so that LDA can be applied. We use both PCA[3] and PLS. The results are quite interesting:

```
> Rand.pcr <- pcr(RandY ~ RandX, ncomp = 2)
> Rand.ldapcr <- lda(RandY ~ scores(Rand.pcr), CV = TRUE)
> table(RandY, Rand.ldapcr$class)

RandY  0   1
    0  5  13
    1  7  15
> Rand.pls <- plsr(RandY ~ RandX, ncomp = 2)
> Rand.ldapls <- lda(RandY ~ scores(Rand.pls), CV = TRUE)
> table(RandY, Rand.ldapls$class)

RandY  0   1
    0 18   0
    1  0  22
```

where the PCA compression leads to results that are reasonably close to the expected 50-50 prediction, PLS-LDA leads to perfect predictions. For random data, that is not exactly what we would want! Note that we did not perform any optimization of the number of latent variables employed—in this case, choosing two latent variables is already enough to be in deep trouble. Validation plots for the regression models *would* have shown that there is trouble ahead—in both the PCR and PLS case, zero latent variables would appear optimal. The moral of the story should by now sound familiar: especially in cases with low ratios of numbers of objects to numbers of variables, one should be very, very careful.

8.7.4 Discussion

Although the concept of using regression methods for classification seems appealing and certainly adds flexibility, there are some remarks that should be made. The question arises what exactly it is that we are predicting. Since the only two reasonable values are zero and one, what do other values signify? How can we compare predictions from different classifiers? If a prediction is far greater than one, does that

---

[3]For simplicity, we employ the pcr function from the **pls** package so that the results can be directly compared with the results from the plsr function.

mean that the classification is more reliable than a prediction that is close to the class cut-off point of 0.5?

More serious are the violations of the usual regression assumptions. In logistic regression, maximum likelihood methods are used to obtain the regression coefficients, but in most cases where a classification is disguised as a regression problem, ordinary least squares methods are used. The assumption of normally distributed errors with equal variance is certainly not fulfilled here. Still, as is so often the case, when a method performs in practice by achieving good quality predictions, people will not be afraid to use it.

Part IV
Model Inspection

Chapter 9
Validation

Validation is the assessment of the quality of a predictive model, in accordance with the scientific paradigm in the natural sciences: a model that is able to make accurate predictions (the position of a planet in two weeks' time) is—in some sense—a "correct" description of reality. In many applications in the natural sciences, unfortunately, validation is hard to do: chemical and biological processes often exhibit quite significant variation unrelated to the model parameters. An example is the circadian rhythm: metabolomic samples, be it from animals or plants, will show very different characteristics when taken at different time points. When the experimental meta-data on the exact time point of sampling are missing, it will be very hard to ascribe differences in metabolite levels to differences between patients and controls, or different varieties of the same plant. Only a rigorous and consistent experimental design will be able to prevent this kind of fluctuations. Moreover, biological variation between individuals often dominates measurement variation. The bigger the variation, the more important it is to have enough samples for validation. Only in this way, reliable error estimates can be obtained.

The main goal usually is to estimate the expected error when applying the model to new, unseen data: the root-mean-square error of prediction (RMSEP). In general, the expected squared error at a point x is given by

$$\text{Err}(x) = E[(Y - \hat{f}(x))^2] \qquad (9.1)$$

which can be decomposed as follows:

$$\text{Err}(x) = (E[\hat{f}(x)] - f(x))^2 + E[(\hat{f}(x) - E[\hat{f}(x)])^2] + \sigma_e^2 \qquad (9.2)$$

where E denotes the usual expectation operator. The first term is the squared bias, corresponding to systematic differences between model predictions and measurements, the second is the variance, and σ_e^2 corresponds to the remaining, irreducible error. In many cases one must strike a balance between bias and variance. Biased regression methods like, e.g., ridge regression and PLS achieve lower MSE values by decreasing the variance component, but pay a price by accepting bias. If it is possible

© Springer-Verlag GmbH Germany, part of Springer Nature 2020
R. Wehrens, *Chemometrics with R*, Use R!,
https://doi.org/10.1007/978-3-662-62027-4_9

to derive confidence intervals for these, this not only provides an idea of the stability of the model, but it can also be useful in determining which variables actually are important in the model.

A second validation aspect, next to estimating the RMSEP, is to assess the size and variability of the model coefficients, summarized here with the term *model stability*. This is especially true for linear models where one can hope to interpret individual coefficients, and perhaps less so for non-linear models. In the example of multiple regression with a singular covariance matrix in Sect. 8.1.1, the variance of the coefficients effectively is infinite, indicating that the model is highly unstable.

Finally, there is the possibility of making use of prior knowledge. Particularly in the natural sciences, one can often assess whether the features that seem important make sense. In a regression model for spectroscopic data, for instance, one would expect wavelengths with large regression coefficients to correspond to peaks in the spectra—large coefficients in areas where no peaks are present would indicate a not too reliable model. Since in most forms of spectroscopy it is possible to associate spectral features with physico-chemical phenomena (specific vibrations, electron transitions, atoms, ...) one can often even say something about the expected sign and magnitude of the regression coefficients. Should these be very different than expected, one may be on to something big—but more likely, one should treat the predictions of such a model with caution, even when the model appears to fit the data well. Typically, more experiments are needed to determine which of the two situations applies. Because of the problem-specific character of this particular type of validation, we will not treat it any further, but will concentrate on the error estimation and model stability aspects.

9.1 Representativity and Independence

One key aspect is that both error estimates and confidence intervals for the model coefficients are derived from the available data (the training data), but that the model will only be relevant when these data are *representative* for the system under study. If there is any systematic difference between the data on which the model is based and the data for which predictions are required, these predictions will be suboptimal and in some cases even horribly wrong. These systematic differences can have several causes: a new machine operator, a new supplier of chemicals or equipment, new schedules of measurement time ("from now on, Saturdays can be used for measuring as well")—all these things may cause new data to be slightly but consistently different from the training data, and as a result the predictive models are no longer optimal. In analytical laboratories, this is a situation that often occurs, and one approach dealing with this is treated in Sect. 11.6.

Especially with extremely large data sets, validation is sometimes based on only one division in a training set and a test set. If the number of samples is very large, the sheer size of the data will usually prevent overfitting and the corresponding error estimates can be quite good. However, it depends on how the training and test sets are constructed. A random division is to be preferred; to be even more sure, several random divisions may be considered. This would also allow one to assess the variability in the validation estimates, and is definitely advisable when computing resources allow it.

One can check whether the training data are really representative for the test data: pathological cases where this is not the case can usually be recognized by simple visualization, (e.g., using PCA). However, one should be very careful not to reject a division too easily: as soon as one starts to use the test data, in this case, to assess whether the division between training and test date is satisfactory, there is the risk of biasing the results. The training set should not only be representative of the test set, but also completely *independent*. An example is the application of the Kennard–Stone algorithm (Kennard and Stone 1969) to make the division in training and test sets. The algorithm selects training samples from the complete data set to cover the complete space of the independent variables as good as possible. However, if the training samples are selected in such a way that they are completely surrounding the test samples, the prediction error on the test set will probably be lower than it should be—it is biased. Of course, when the algorithm is only used to decrease the number of samples in the training set, and the test set has been set aside before the Kennard–Stone algorithm is run, then there is no problem (provided the discarded training-set samples are not added to the test set!) and we can still treat the error on the test set as an unbiased estimate of what we can expect for future samples.

If the available data can be assumed to be representative of the future data, we can use them in several ways to assess the quality of the predictions. The main point in all cases is the same: from the data at hand, we simulate a situation where unseen data have to be predicted. In *crossvalidation*, this is done by leaving out part of the data, and building the model on the remainder. In *bootstrapping*, the other main validation technique, the data are resampled with replacement, so that some data points are present several times in the training set, and others (the "out-of-bag", or OOB, samples) are absent. The performance of the model(s) on the OOB samples is then an indication of the prediction quality of the model for future samples.

In estimating errors, one should take care not to use *any* information of the test set: if the independence of training and test sets is compromised error estimates become biased. An often-made error is to scale (autoscaling, mean-centering) the data before the split into training and test sets. Obviously, the information of the objects in the test set *is* being used: column means and standard deviations are influenced by data from the test set. This leads to biased error estimates—they are, in general, lower that they should be. In the crossvalidation routines of the **pls** package, for example, scaling of the data is done in the correct way: the OOB samples in a crossvalidation iteration are scaled using the means (and perhaps variances) of the in-bag samples. If, however, other forms of scaling are necessary, this can not be done automatically. The **pls** package provides an explicit `crossval` function, which makes it possible to include sample-specific scaling functions in the calling formula:

```
> gasoline.mscpcr <- pcr(octane ~ msc(NIR), data = gasoline,
+                        ncomp = 4)
> gasoline.mscpcr.cv <- crossval(gasoline.mscpcr, length.seg = 1)
> RMSEP(gasoline.mscpcr.cv, estimate = "CV")
(Intercept)      1 comps      2 comps      3 comps      4 comps
     1.5430       1.4589       0.8901       0.2598       0.2668
```

This particular piece of code applies multiplicative scatter correction (MSC, see Sect. 3.2) on all in-bag samples, and scales the OOB samples in the same way, as it should be done. Interestingly, this leads to a PCR model where three components would be optimal, one fewer component than without the MSC scaling.

A final remark concerns more complicated experimental designs. The general rule is that the design should be taken into account when setting up the validation. As an example, consider a longitudinal experiment where multiple measurements of the same objects at different time points are present in the data. When applying subsampling approaches like crossvalidation to such data sets one should leave out complete objects, rather than individual measurements: obviously multiple measurements of the same object, even taken at different times, are not independent. Randomly sampling individual data points would probably lead to over-optimistic validation estimates.

9.2   Error Measures

A distinction has to be made between the prediction of a continuous variable (regression), and a categorical variable, as in classification. In regression, the root-mean-square error of validation (RMSEV) is given, analogously to Eq. 8.12, by

$$\text{RMSEV} = \sqrt{\frac{\sum_i (\hat{y}_{(i)} - y_{(i)})^2}{n}} \tag{9.3}$$

where $y_{(i)}$ is the out-of-bag sample in a crossvalidation or bootstrap. That is, the predictions are made for samples that have not been used in building the model. A summary of these prediction errors can be used as an estimate for future performance. In this case, the average of the sum of squared errors is taken—sometimes there are better alternatives.

For classification, the simplest possibility is to look at the fraction of correctly classified observations. in R:

```
> err.rate <- function(x, y) sum(x != y) / length(x)
```

A more elaborate alternative is to assign each type of misclassification a *cost*, and to minimize a loss function consisting of the total costs associated with misclassifications. In a two-class situation, for example, this makes it possible to prevent false negatives at the expense of accepting more false positives; in a medical context, it may be the case that a specific test should recognize all patients with a specific disease, even if that means that a few people without the disease are also tagged. Missing a positive sample (a false negative outcome) in this example has much more radical consequences than the reverse, incorrectly calling a healthy person ill.

A related alternative is to focus on the two components of classification accuracy, *sensitivity* and *specificity*. Sensitivity, also known as the *recall rate* or the *true positive rate*, is the fraction of objects from a particular class k which are actually assigned to that class:

$$\text{sensitivity}_k = \frac{TP_k}{TP_k + FN_k} \tag{9.4}$$

where TP_k is the number of True Positives (i.e., objects correctly assigned to class k) and FN_k is the number of False Negatives (objects belonging to class k but classified otherwise). A sensitivity of one indicates that all objects of class k are assigned to the correct class—note that many other objects, not of class k, may be assigned to that class as well.

Specificity is related to the purity of class predictions, and summarizes the fraction of objects in class k that belong elsewhere:

$$\text{specificity}_k = \frac{TN_k}{FP_k + TN_k} \tag{9.5}$$

TN_k and FP_k indicate True Negatives and False Positives for class k, respectively. A specificity of one indicates that no objects have been classified as class k incorrectly. The measure 1—specificity is sometimes referred to as the *false positive rate*.

In practice, one will have to compromise between specificity and sensitivity: usually, sensitivity can be increased at the expense of specificity and vice versa by changing parameters of the classification procedure. For two-class problems, a common visualization is the Receiver Operating Characteristic (ROC, Brown and Davis 2006), which plots the true positive rate against the false positive rate for several values of the classifier threshold. Consider, e.g., the optimization of k, the number of neighbors in the KNN classification of the wine data. Let us focus on the distinction between Barbera and Grignolino, where we (arbitrarily) choose Barbera as the positive class, and Grignolino as negative.

```
> X <- wines[vintages != "Barolo", ]
> vint <- factor(vintages[vintages != "Barolo"])
> kvalues <- 1:12
> ktabs <- lapply(kvalues,
+                 function(i) {
+                     kpred <- knn.cv(X, vint, k = i)
+                     table(vint, kpred)
+                 })
```

For twelve different values of k we calculate the crossvalidated predictions and we save the crosstable. From the resulting list we can easily calculate true positive and false positive rates:

Fig. 9.1 ROC curve
(zoomed in to display only
the relative part) for the
discrimination between
Grignolino and Barbera
wines using different values
of *k* in KNN classification.
Predictions are
LOO-crossvalidated

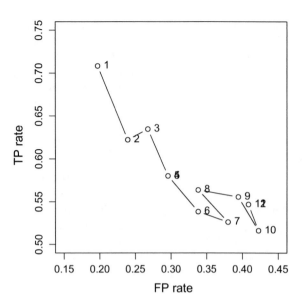

```
> TPrates <- sapply(ktabs, function(x) x[1, 1]/sum(x[, 1]))
> FPrates <- sapply(ktabs, function(x) 1 - x[2, 2]/sum(x[2, ]))
> plot(FPrates, TPrates, type = "b",
+       xlim = c(.15, .45), ylim = c(.5, .75),
+       xlab = "FP rate", ylab = "TP rate")
> text(FPrates, TPrates, 1:12, pos = 4)
```

In this case, the result, shown in Fig. 9.1, leaves no doubt that $k = 1$ gives the best
results: it shows the lowest fraction of false positives (i.e., Grignolinos predicted as
Barberas) as well as the highest fraction of true positives. The closer a point is to the
top left corner (perfect prediction), the better.

Note that a careful inspection of model residuals should be a standard ingredient
of any analysis. Just summing up the number of misclassifications, or squared errors,
is not telling the whole story. In some parts of the data space one might see, e.g., more
misclassifications or larger errors than in other parts. For simple univariate regression,
standard plots exist (simply plotting an lm object in R will give a reasonable subset)—
for multivariate models techniques like PCA can come in handy, but there is ample
opportunity for creativity from the part of the data analyst.

9.3 Model Selection

In the field of *model selection*, one aims at selecting the best model amongst a series
of possibilities, usually based on some quality criterion such as an error estimate.
What makes model selection slightly special is that we are not interested in the error

estimates themselves, but rather in the order of the sizes of the errors: we would like to pick the model with the smallest error. Also biased error estimates are perfectly acceptable when the bias does not influence our choice, and in some cases biased estimates are even preferable since they often have a lower variance. We will come back to this in later sections, discussing different resampling-based estimates.

9.3.1 Permutation Approaches

A form of resampling that we have not yet touched upon is *permutation*, randomly redistributing labels in order to simulate one possible realization of the data under a null hypothesis. The concept is most easily explained in the context of classification. Suppose we have a classifier that distinguishes between two classes, A and B, each represented by the same number of samples. Let's say the classifier achieves a 65% correct prediction rate. The question is whether this could be due to chance. In a permutation test, one would train the same classifier many times on a data set in which class labels A and B would be randomly assigned to samples. Since in such a permutation there is no relation between the dependent and independent variables, one would expect a success rate of 50%. In practice one will see variation. Comparing the prediction rate observed with the real data with the quantiles of the permutation prediction rates gives an estimate of the p value, or in other words, tells you whether the model is significant or not. In the example above: if we would do 500 permutations and in 78 of them we would find prediction rates above 65%, we should conclude that our classifier is not doing significantly better than a chance process. If only three of the 500 permutations would lead to prediction rates of 65% or more, on the other hand, we would declare our model significant.

So where other forms of validation try to obtain an error measure, permutation testing as described above aims to assess significance. While the principle remains the same, there are other ways in which the permutation test can be used. One example was given in Sect. 8.2.2 in the context of establishing the optimal number of components in a PCR or PLS regression model. There, residuals of models using A and $A + 1$ components, respectively, are being permuted and the true sum of squares is then compared to the null distribution given by the set of permutations. If there is no significant difference between the two, the smallest model with A components is preferred. In the remainder of this chapter we'll focus on establishing estimates for the magnitude of the prediction errors.

9.3.2 Model Selection Indices

Resampling approaches such as crossvalidation can be time-consuming, especially for large data sets or complicated models. In such cases simple, direct estimates could form a valuable alternative. The most common ones consist of a term indicating the agreement between empirical and predicted values, and a penalty for model complexity. Important examples are Mallows's C_p (Mallows 1973) and the AIC and BIC values (Akaike 1974; Schwarz 1978), already encountered in Sect. 6.3. The C_p value is a special case of AIC for general models, adjusting the expected value in such a way that it is approximately equal to the prediction error. In a regression context, these two measures are equal, given by

$$\text{AIC} = C_p = \text{MSE} + 2 \times p\, \hat{\sigma}^2/n \tag{9.6}$$
$$\text{BIC} = \text{MSE} + \log n \times p\, \hat{\sigma}^2/n \tag{9.7}$$

where n is the number of objects, p is the number of parameters in the model, MSE is the mean squared error of calibration, and $\hat{\sigma}^2$ is an estimate of the residual variance—an obvious choice would be $\text{MSE}/(n - p)$ (Efron and Tibshirani 1993). It can be seen that, for any practical data size, BIC penalizes more heavily than C_p and AIC, and therefore will choose more parsimonious models. For model selection in the life sciences, these statistics have never really been very popular. A simple reason is that it is hard to assess the "true" value of p: how many degrees of freedom do you have in a PLS or PCR regression? Methods like crossvalidation are more simple to apply and interpret—and with computing power being cheap, scientists happily accept the extra computational effort associated with it.

9.3.3 Including Model Selection in the Validation

Up to now we have concentrated on validation approaches such as crossvalidation for particular models, e.g., a PLS model with four components, or a KNN classifier with $k = 3$. Typically, we would repeat this validation for other numbers of latent variable, or other values of k, and base the selection of the best model on some kind of decision rule (e.g., the approaches mentioned in Sect. 8.2.2 for choosing the number of latent variables in a multivariate regression model). As has been stated before, the CV error estimate associated with the selected model is to be interpreted in a relative way, indicating which of the models under comparison is the best one—it's value should not be taken absolutely. The reason is the model selection process: we choose this model precisely *because* it has the lowest error, and so we introduce a downward bias.

There is another option. Rather than performing crossvalidation (or bootstrapping or any other validation technique) on one fully specified model, one could also use it on the complete procedure, including the model selection (Efron and Hastie 2016). That is, if we decide to choose the optimal number of latent variables in a PLS model using the one-sigma rule mentioned in Sect. 8.2.2, we could simply apply crossvalidation on the overall procedure, including applying the selection rule. The resulting error estimate[1] now *is* an unbiased estimate of what we can expect for future data. Note that in each crossvalidation iteration the optimal number of components may be different. This usually adds variance, so error estimates obtained with this procedure are expected to be larger than the estimates we have been discussing until now, which makes sense.

9.4 Crossvalidation Revisited

Crossvalidation, as we already have seen, is a simple and trustworthy method to estimate prediction errors. There are two main disadvantages of LOO crossvalidation. The first is the time needed to perform the calculations. Especially for data sets with many objects and time-consuming modelling methods, LOO may be too expensive to be practical. There are two ways around this problem: the first is to use fast alternatives to direct calculations—in some cases analytical solutions exist, or fast and good approximations. A second possibility is to focus on leaving out larger segments at a time. This latter option also alleviates the second disadvantage of LOO crossvalidation—the relatively large variability of its error estimates.

9.4.1 *LOO Crossvalidation*

Let us once again look at the equation for the LOO crossvalidation error:

$$\varepsilon_{CV}^2 = \frac{1}{n} \sum_{i=1}^{n} \left(y_{(i)} - \hat{y}_{(i)} \right)^2 = \frac{1}{n} \sum_{i=1}^{n} \varepsilon_{(i)}^2 \tag{9.8}$$

where subscript (i) indicates that observation i is being predicted while not being part of the training data. Although the procedure is simple to understand and implement, it can take a lot of time to run for larger data sets. However, for many modelling methods it is not necessary to calculate the n different models explicitly. For ordinary least-squares regression, for example, one can show that the ith residual of a LOO crossvalidation is given by

$$\varepsilon_{(i)}^2 = \varepsilon_i^2 / (1 - h_{ii}) \tag{9.9}$$

where ε_i^2 is the squared residual of sample i when it is *included* in the training set, and h_{ii} is the ith diagonal element of the hat matrix H, given by

$$H = X \left(X^T X \right)^{-1} X^T \tag{9.10}$$

---

[1]This in effect is an example of double crossvalidation, since the selection rule internally uses crossvalidation, too. We'll come back to this in a later section in this chapter.

Therefore, the LOO error estimate can be obtained without explicit iteration by

$$\varepsilon_{CV}^2 = \frac{1}{n} \sum_{i=1}^{n} \left(\frac{y_i - \hat{y}_i}{1 - h_{ii}} \right)^2 \tag{9.11}$$

This shortcut is available in all cases where it is possible to write the predicted values as a product of a type of hat matrix $\boldsymbol{H}$, independent of y, and the measured y values:

$$\hat{y} = \boldsymbol{H} y \tag{9.12}$$

Generalized crossvalidation (GCV, Craven and Wahba 1979) goes one step further: instead of using the individual diagonal elements of the hat matrix h_{ii}, the average diagonal element is used:

$$\varepsilon_{GCV}^2 = \frac{1}{n \left(1 - \sum_{j=1}^{n} h_{jj} \right)^2} \sum_{i=1}^{n} \left(y_i - \hat{y}_i \right)^2 \tag{9.13}$$

Applying these equations to PCR leads to small differences with the usual LOO estimates, since the principal components that are estimated when leaving out each sample in turn will deviate slightly (assuming there are no gross outliers). Consider the (bad) fit of the one-component PCR model for the gasoline data, calculated with explicit construction of n sets of size $n - 1$:

```
> gasoline.pcr <- pcr(octane ~ ., data = gasoline,
+                     validation = "LOO", ncomp = 1)
> RMSEP(gasoline.pcr, estimate = "CV")
(Intercept)      1 comps
      1.543        1.447
```

The estimate based on Eq. 9.11 is obtained by

```
> gasoline.pcr2 <- pcr(octane ~ ., data = gasoline, ncomp = 1)
> X <- gasoline.pcr2$scores
> HatM <- X %*% solve(crossprod(X), t(X))
> sqrt(mean((gasoline.pcr2$residuals/(1 - diag(HatM)))^2))
[1] 1.4187
```

The GCV estimate from Eq. 9.13 deviates more from the LOO result:

```
> sqrt(mean((gasoline.pcr2$residuals/(1 - mean(diag(HatM))))^2))
[1] 1.3888
```

If one is willing to ignore the variation in the PCs introduced by leaving out individual objects, as may be perfectly acceptable in the case of data sets with many objects, this provides a way to significantly speed up calculations. The example above was four times faster than the explicit loop, as is implemented in the pcr function with the validation = "LOO" argument. For PLS, it is a different story: there, the latent variables are estimated using y, and Eq. 9.12 does not hold.

9.4.2 Leave-Multiple-Out Crossvalidation

Instead of leaving out one sample at a time, it is also possible to leave out a sizeable fraction, usually 10% of the data; the latter is also called "ten-fold crossvalidation". This approach has become quite popular—not only is it roughly ten times faster, it also shows less variability in the error estimates (Efron and Tibshirani 1993). Again, there is a bias-variance trade-off: the variance may be smaller, but a small bias occurs because the model is based on a data set that is appreciably smaller than the "real" data set, and therefore is slightly pessimistic by nature.

This "leave-multiple-out" (LMO) crossvalidation is usually implemented in a random way: the order of the rows of the data matrix is randomized, and consecutive chunks of roughly equal size are used as test sets. In case the data are structured, it is possible to use non-randomized chunks: the functions in the **pls** package have special provisions for this. The following lines of code lead, e.g., to interleaved sample selection:

```
> gasoline.pcr <- pcr(octane ~ ., data = gasoline,
+                     validation = "CV", ncomp = 4,
+                     segment.type = "interleaved")
> RMSEP(gasoline.pcr, estimate = "CV")
(Intercept)    1 comps    2 comps    3 comps    4 comps
     1.5430     1.4261     1.4457     1.2179     0.2468
```

An alternative is to use `segment.type = "consecutive"`. Also, it is possible to construct the segments (i.e., the crossvalidation sets) by hand or otherwise, and explicitly present them to the modelling function using the `segments` argument. See the manual pages for more information.

9.4.3 Double Crossvalidation

In all cases where crossvalidation is used to establish optimal values for modelling parameters, the resulting error estimates are not indicative of the performance of future observations. They are biased, in that they are used to pick the optimal model. Another round of validation is required. This leads to *double crossvalidation* (Stone 1974), as visualized in Fig. 9.2: the inner crossvalidation loop is used to determine the optimal model parameters, very often, in chemometrics, the optimal number of latent variables, and the outer crossvalidation loop assesses the corresponding prediction error. At the expense of more computing time, one is able to select optimal model parameters as well as estimate prediction error.

The problem is that usually one ends up selecting different parameter settings in different crossvalidation iterations: leaving out segment 1 may lead to a PLS model with two components, whereas segment two may seem to need four PLS components. Which do you choose? Averaging is no solution—again, one would be using information which is not supposed to be available, and the resulting error

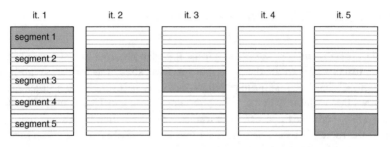

Fig. 9.2 Double crossvalidation: the inner CV loop, indicated by the gray horizontal lines, is used to estimate the optimal parameters for the modelling method. The outer loop, a five-fold crossvalidation, visualized by the gray rectangles, is used to estimate the prediction error

estimates would be biased. One approach is to use all optimal models simultaneously, and average the predictions (Smit et al. 2007). The disadvantage is that one loses the interpretation of one single model; however, this may be a reasonable price to pay. Other so-called ensemble methods will be treated in Sects. 9.7.1 and 9.7.2.

9.5 The Jackknife

Jackknifing (Efron and Tibshirani 1993) is the application of crossvalidation to obtain statistics other than error estimates, usually pertaining to model coefficients. The jackknife can be used to assess the bias and variance of regression coefficients. The jackknife estimate of bias, for example, is given by

$$\widehat{\mathrm{Bias}}_{jck}(b) = (n-1)(\bar{b}_{(i)} - b) \tag{9.14}$$

where b is the regression coefficient[2] obtained with the full data, and $b_{(i)}$ is the coefficient from the data with sample i removed, just like in LOO crossvalidation. The bias estimate is simply the difference between the average of all these LOO estimates, and the full-sample estimate, multiplied by the factor $n - 1$.

Let us check the bias of the PLS estimates on the gasoline data using two latent variables. The `plsr` function, when given the argument `jackknife = TRUE`,[3] is keeping all regression coefficients of a LOO crossvalidation in the `validation` element of the fitted object, so finding the bias estimates is not too difficult:

---

[2]In a multivariate setting we should use an index such as b_j—to avoid complicated notation we skip that for the moment.

[3]Information on this functionality can be found in the manual page of function `mvrCv`.

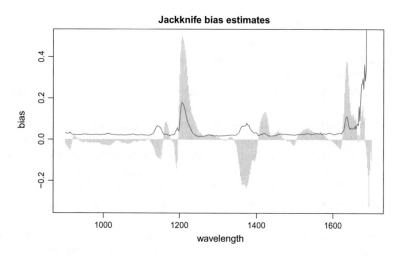

Fig. 9.3 Jackknife estimates of bias (in gray) and variance (red line) for a two-component PLS model on the gasoline data

```
> gasoline.pls <- plsr(octane ~ ., data = gasoline,
+                         validation = "LOO", ncomp = 2,
+                         jackknife = TRUE)
> n <- length(gasoline$octane)
> b.oob <- gasoline.pls$validation$coefficients[, , 2, ]
> bias.est <- (n-1) * (rowMeans(b.oob) - coef(gasoline.pls))
> plot(wavelengths, bias.est, xlab = "wavelength", ylab = "bias",
+       type = "h", main = "Jackknife bias estimates",
+       col = "gray")
```

The result is shown in Fig. 9.3—clearly, the bias for specific coefficients can be appreciable.

The jackknife estimate of variance is given by

$$\widehat{\text{Var}}_{jck}(b) = \frac{n-1}{n} \sum (b_{(i)} - \bar{b}_{(i)})^2 \tag{9.15}$$

and is implemented in `var.jack`. Again, an object of class `mvr` needs to be supplied that is fitted with `jackknife = TRUE`:

```
> var.est <- var.jack(gasoline.pls)
> lines(wavelengths, var.est, col = "red")
```

The variance is shown as the red line in Fig. 9.3. In the most important regions, bias seems to dominate variance.

Several variants of the jackknife exist, including some where more than one sample is left out (Efron and Tibshirani 1993). In practice, however, the jackknife has been replaced by the more versatile bootstrap.

9.6 The Bootstrap

The bootstrap (Efron and Tibshirani 1993; Davison and Hinkley 1997) is a generalization of the ideas behind crossvalidation: again, the idea is to generate multiple data sets that, after analysis, shed light on the variability of the statistic of interest as a result of the different training set compositions. Rather than splitting up the data to obtain training and test sets, in *non-parametric bootstrapping* one generates a training set—a bootstrap sample—by sampling with replacement from the data. Whereas the measured data set is one possible realization of the underlying population, an individual bootstrap sample is, analogously, one realization from the complete set. Since we may have sufficient knowledge of difference between the complete set and the empirical realizations, simply by generating more bootstrap samples, we can study the distribution of the statistic of interest θ. In non-parametric bootstrapping applied to regression problems, there are two main approaches for generating a bootstrap sample. One is to sample (again, with replacement) from the *errors* of the initial model. Bootstrap samples are generated by adding the resampled errors to the original data. This strategy is appropriate when the X data can be regarded as fixed and the model is assumed to be correct. In other cases, one can sample complete cases, i.e., rows from the data matrix, to obtain a bootstrap sample. In such a bootstrap sample, some rows are present multiple times; others are absent.

In *parametric bootstrapping* on the other hand, one describes the data with a parametric distribution, from which then random bootstrap samples are generated. In the life sciences, high-dimensional data are the rule rather than the exception, and therefore any parametric description of a data set is apt to be based on very sparse data. Consequently, the parametric bootstrap has been less popular in this context.

What method is used to generate the bootstrap distribution, parametric bootstrapping or non-parametric bootstrapping, is basically irrelevant for the subsequent analysis. Typically, several hundreds to thousands bootstrap samples are analyzed, and the variability of the statistic of interest is monitored. This enables one to make inferences, both with respect to estimating prediction errors and confidence intervals for model coefficients.

9.6.1 Error Estimation with the Bootstrap

Because a bootstrap sample will effectively never contain all samples in the data set, there are samples that have not been involved in building the model. These out-of-bag samples can conveniently be used in estimation of prediction errors. A popular estimator is the so-called 0.632 estimate $\hat{\varepsilon}_{0.632}$, given by

$$\hat{\varepsilon}_{0.632}^2 = 0.368 \, \text{MSEC} + 0.632 \bar{\varepsilon}_B^2 \tag{9.16}$$

where $\bar{\varepsilon}_B^2$ is the average squared prediction error of the OOB samples in the B bootstrap samples, and MSEC is the mean squared training error (on the complete data set). The factor $0.632 \approx (1 - e^{-1})$ is approximately the probability of a sample to end up in a bootstrap sample (Efron and Tibshirani 1993). In practice, the 0.632 estimator is the most popular form for estimating prediction errors; a more sophisticated version, correcting possible bias, is known as the 0.632+ estimator (Efron and Tibshirani 1997) but in many cases the difference is small.

As an example, let us use bootstrapping rather than crossvalidation to determine the optimal number of latent variables in PCR fitting of the gasoline data. In this case, the independent variables are not fixed, and there is some uncertainty on whether the model is correct. This leads to the adoption of the resampling cases paradigm. We start by defining bootstrap sample indices—in this case we take 500 bootstrap samples.

```
> B <- 500
> ngas <- nrow(gasoline)
> boot.indices <-
+    matrix(sample(1:ngas, ngas * B, replace = TRUE), ncol = B)
> sort(boot.indices[, 1])[1:20]
 [1]  2  2  3  3  4  6  8  8  8  8 11 12 14 15 15 16 17 19 20 21
```

Among others, objects 1 and 5 are absent from the first bootstrap sample, (partially) shown here as an example. Other samples, such as 2 and 3, occur multiple times. Similar behaviour is observed for the other 499 bootstrap samples. We now build a PCR model for each bootstrap sample and record the predictions of the out-of-bag objects. The following code is not particularly memory-efficient but easy to understand:

```
> npc <- 5
> predictions <- array(NA, c(ngas, npc, B))
> for (i in 1:B) {
+    gas.bootpcr <- pcr(octane ~ ., data = gasoline,
+                       ncomp = npc, subset = boot.indices[, i])
+    oobs <- (1:ngas)[-boot.indices[, i]]
+    predictions[oobs, , i] <-
+      predict(gas.bootpcr,
+              newdata = gasoline$NIR[oobs, ])[, 1, ]
+ }
```

Next, the OOB errors for the individual objects are calculated, and summarized in one estimate:

```
> diffs <- sweep(predictions, 1, gasoline$octane)
> sqerrors <- apply(diffs^2, c(1, 2), mean, na.rm = TRUE)
> sqrt(colMeans(sqerrors))
[1] 1.48695 1.50077 1.24562 0.28667 0.27598
```

Finally, the out-of-bag errors are combined with the calibration error to obtain the 0.632 estimate:

```
> gas.pcr <- pcr(octane ~ ., data = gasoline, ncomp = npc)
> RMSEP(gas.pcr, intercept = FALSE)
1 comps   2 comps   3 comps   4 comps   5 comps
 1.3656    1.3603    1.1097    0.2305    0.2260
> error.632 <- .368 * colMeans(gas.pcr$residuals^2) +
+    .632 * colMeans(sqerrors)
> sqrt(error.632)
          1 comps 2 comps 3 comps 4 comps 5 comps
octane    1.4435  1.4507  1.1974 0.26737 0.25873
```

The result is an upward correction of the too optimistic training set errors. We can
compare the 0.632 estimate with the LOO and ten-fold crossvalidation estimates:

```
> gas.pcr.cv <- pcr(octane ~ ., data = gasoline, ncomp = npc,
+                    validation = "CV")
> gas.pcr.loo <- pcr(octane ~ ., data = gasoline, ncomp = npc,
+                    validation = "LOO")
> bp <- barplot(sqrt(error.632),
+                ylim = c(0, 1.6), col = "peachpuff")
> lines(bp, sqrt(c(gas.pcr.cv$validation$PRESS) / ngas),
+        col = 2, lwd = 2)
> lines(bp, sqrt(c(gas.pcr.loo$validation$PRESS) / ngas),
+        col = 4, lty = 2, lwd = 2)
> legend("topright", lty = 1:2, col = c(2, 4), lwd = 2,
+        legend = c("CV", "LOO"))
```

The result is shown in Fig. 9.4. The estimates in general agree very well—the differences that can be seen are the consequence of the stochastic nature of both ten-
fold crossvalidation and bootstrapping: every time a slightly different result will be
obtained.

Fig. 9.4 Error estimates for
PCR on the gasoline data:
bars indicate the result of the
0.632 bootstrap, the solid
line is the ten-fold
crossvalidation, and the
dashed line the LOO
crossvalidation

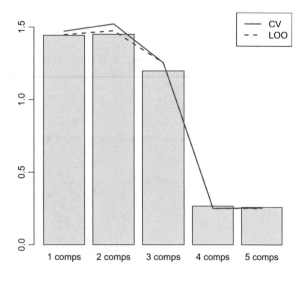

It now should be clear what is the philosophy behind the 0.632 estimator. What it estimates, in fact, is the amount of optimism associated with the RMSEC value, $\hat{\omega}_{0.632}$:

$$\hat{\omega}_{0.632} = 0.632(\text{MSEC} - \bar{\varepsilon}_B) \tag{9.17}$$

The original estimate is then corrected for this optimism:

$$\hat{\varepsilon}_{0.632} = \text{MSEC} + \hat{\omega}_{0.632} \tag{9.18}$$

which leads to Eq. 9.16.

Several R packages are available that contain functions for bootstrapping. Perhaps the two best known ones are **bootstrap**, associated with Efron and Tibshirani (1993), and **boot**, written by Angelo Canty and implementing functions from Davison and Hinkley (1997). The former is a relatively simple package, maintained mostly to support Efron and Tibshirani (1993)—**boot**, a recommended package, is the primary general implementation of bootstrapping in R. The implementation of the 0.632 estimator using **boot** is done in a couple of steps (Davison and Hinkley 1997, p. 324). First, the bootstrap samples are generated, returning the statistic to be bootstrapped—in this case, the prediction errors[4]:

```
> gas.pcr.boot632 <-
+   boot(gasoline,
+        function(x, ind) {
+          mod <- pcr(octane ~ ., data = x,
+                     subset = ind, ncomp = 4)
+          gasoline$octane -
+            predict(mod, newdata = gasoline$NIR, ncomp = 4)},
+        R = 499)
```

The optimism is assessed by only considering the errors of the out-of-bag samples. For every bootstrap sample, we can find out which samples are constituting it using the boot.array function:

```
> dim(boot.array(gas.pcr.boot632))
[1] 499  60
> boot.array(gas.pcr.boot632)[1, 1:10]
 [1] 0 1 0 1 2 1 0 1 2 1
```

Just like when we did the resampling ourselves, some objects are absent from this bootstrap sample (here, as an example, using the first, only showing the first ten objects), and others are present multiple times. Averaging the squared errors of the OOB objects leads to the 0.632 estimate:

---

[4]In Davison and Hinkley (1997) and the corresponding **boot** package the number of bootstrap samples is typically a number like 499 or 999—the original sample then is added to the bootstrap set. Most other implementations use 500 and 1000. The differences are not very important in practice.

```
> in.bag <- boot.array(gas.pcr.boot632)
> oob.error <- mean((gas.pcr.boot632$t^2)[in.bag == 0])
> app.error <- MSEP(pcr(octane ~ ., data = gasoline, ncomp = 4),
+                    ncomp = 4, intercept = FALSE)
> sqrt(.368 * c(app.error$val) + .632 * oob.error)
[1] 0.26572
```

This error estimate is very similar to the four-fold crossvalidation result in Sect. 9.4.2 (0.2468). Note that it is not exactly equal to the 0.632 estimate in Sect. 9.6.1 (0.26737) because different bootstrap samples have been selected, but again the difference is small.

9.6.2   Confidence Intervals for Regression Coefficients

The bootstrap may also be used to assess the variability of a statistic such as an error estimate. A particularly important application in chemometrics is the standard error of a regression coefficient from a PCR or PLS model. Alternatively, confidence intervals can be built for the regression coefficients. No analytical solutions such as those for MLR exist in these cases; nevertheless, we would like to be able to say something about which coefficients are actually contributing to the regression model.

Typically, for an interval estimate such as a confidence interval, more bootstrap samples are needed than for a point estimate, such as an error estimate. Several hundred bootstrap samples are taken to be sufficient for point estimates; several thousand for confidence intervals. Taking smaller numbers may drastically increase the variability of the estimates, and with the current abundance of computing power there is rarely a case for being too economical.

The simplest possible approach is the *percentile* method: estimate the models for B bootstrap samples, and use the $B\alpha/2$ and $B(1 - \alpha/2)$ values as the $(1 - \alpha)$ confidence intervals. For the gasoline data, modelled with PCR using four PCs, these bootstrap regression coefficients are obtained by:

```
> B <- 1000
> ngas <- nrow(gasoline)
> boot.indices <-
+   matrix(sample(1:ngas, ngas * B, replace = TRUE), ncol = B)
> npc <- 4
> gas.pcr <- pcr(octane ~ ., data = gasoline, ncomp = npc)
> coefs <- matrix(0, ncol(gasoline$NIR), B)
> for (i in 1:B) {
+   gas.bootpcr <- pcr(octane ~ ., data = gasoline,
+                      ncomp = npc, subset = boot.indices[, i])
+   coefs[, i] <- c(coef(gas.bootpcr))
+ }
```

A plot of the area covered by the regression coefficients of all bootstrap samples is shown in Fig. 9.5:

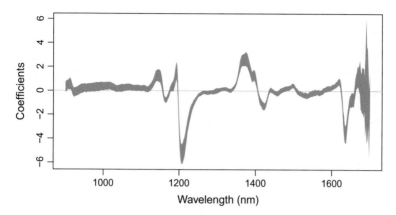

Fig. 9.5 Regression coefficients from all 1000 bootstrap samples for the gasoline data, using PCR with four latent variables

```
> matplot(wavelengths, coefs, type = "n",
+         ylab = "Coefficients", xlab = "Wavelength (nm)")
> abline(h = 0, col = "gray")
> polygon(c(wavelengths, rev(wavelengths)),
+         c(apply(coefs, 1, max), rev(apply(coefs, 1, min))),
+         col = "steelblue", border = NA)
```

Some of the wavelengths show considerable variation in their regression coefficients, especially the longer wavelengths above 1650 nm.

In the percentile method using 1000 bootstrap samples, the 95% confidence intervals are given by the 25th and 975th ordered values of each coefficient:

```
> coef.stats <- cbind(apply(coefs, 1, quantile, .025),
+                     apply(coefs, 1, quantile, .975))
> matplot(wavelengths, coef.stats, type = "n",
+         xlab = "Wavelength (nm)",
+         ylab = "Regression coefficient")
> abline(h = 0, col = "gray")
> polygon(c(wavelengths, rev(wavelengths)),
+         c(coef.stats[, 1], rev(coef.stats[, 2])),
+         col = "pink", border = NA)
> lines(wavelengths, c(coef(gas.pcr)))
```

The corresponding plot is shown in Fig. 9.6. Since the most extreme values will be removed by the percentile strategy, these CIs are more narrow than the area covered by the bootstrap coefficients from Fig. 9.5. Clearly, for most coefficients, zero is not in the confidence interval. A clear exception is seen in the longer wavelengths: there, the confidence intervals are very wide, indicating that this region contains very little relevant information.

The percentile method was the first attempt at deriving confidence intervals from bootstrap samples (Efron 1979) and has enjoyed huge popularity; however, one can show that the intervals are, in fact, incorrect. If the intervals are not symmetric (and it

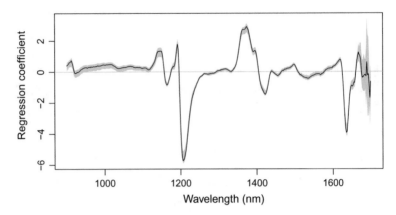

Fig. 9.6 Regression vector and 95% confidence intervals for the individual coefficients, for the PCR model of the gasoline data with four PCs. Confidence intervals are obtained with the bootstrap percentile method

can be seen in Fig. 9.6 that this is quite often the case—it is one of the big advantages of bootstrapping methods that they are able to define asymmetric intervals), it can be shown that the percentile method uses the skewness of the distribution the wrong way around (Efron and Tibshirani 1993). Better results are obtained by so-called *studentized* confidence intervals, in which the statistic of interest is given by

$$t_b = \frac{\hat{\theta}_b - \hat{\theta}}{\hat{\sigma}_b} \tag{9.19}$$

where $\hat{\theta}_b$ is the estimate for the statistic of interest, obtained from the bth bootstrap sample, $\hat{\sigma}_b$ is the standard deviation of that estimate, and $\hat{\theta}$ is the estimate obtained from the complete original data set. In the example of regression, $\hat{\theta}$ corresponds to the regression coefficient at a certain wavelength. Often, no analytical expression exists for $\hat{\sigma}_b$, and it should be obtained by other means, e.g., crossvalidation, or an inner bootstrap loop. Using the notation of $t_{B\alpha/2}$ as an approximation for the $\alpha/2$th quantile of the distribution of t_b, the studentized confidence intervals are given by

$$\hat{\theta} - t_{B(1-\alpha/2)} \leq \theta \leq \hat{\theta} - t_{B\alpha/2} \tag{9.20}$$

Several other ways of estimating confidence intervals exist, most notably the bias-corrected and accelerated (BCα) interval (Efron and Tibshirani 1993; Davison and Hinkley 1997).

The **boot** package provides the function `boot.ci`, which calculates several confidence interval estimates in one go. Again, first the bootstrap sampling is done and the statistics of interest are calculated:

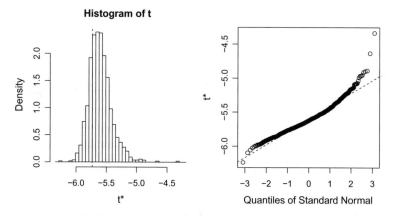

Fig. 9.7 Bootstrap plot for the regression coefficient at 1206 nm; in all bootstrap samples the coefficient is much smaller than zero

```
> gas.pcr.bootCI <-
+    boot(gasoline,
+         function(x, ind) {
+             c(coef(pcr(octane ~ ., data = x,
+                        ncomp = npc, subset = ind)))},
+         R = 999)
```

Here we use R = 999 to conform to the setup of the **boot** package—the actual sample is seen as the 1000th element of the set. The regression coefficients are stored in the gas.pcr.bootCI object, which is of class "boot", in the element named t:

```
> dim(gas.pcr.bootCI$t)
[1] 999 401
```

Plots of individual estimates can be made through the index argument:

```
> smallest <- which.min(gas.pcr.bootCI$t0)
> plot(gas.pcr.bootCI, index = smallest)
```

From the plot, shown in Fig. 9.7, one can see the distribution of the values for this coefficient in all bootstrap samples—the corresponding confidence interval will definitely not contain zero. The dashed line indicates the estimate based on the full data; these estimates are stored in the list element t0.

Confidence intervals for individual coefficients can be obtained from the gas.pcr.bootCI object as follows:

```
> boot.ci(gas.pcr.bootCI, index = smallest, type = c("perc", "bca"))
BOOTSTRAP CONFIDENCE INTERVAL CALCULATIONS
Based on 999 bootstrap replicates

CALL :
boot.ci(boot.out = gas.pcr.bootCI, type = c("perc", "bca"),
        index = smallest)

Intervals :
Level      Percentile                  BCa
95%     (-5.909, -5.157 )      (-6.239, -5.538 )
Calculations and Intervals on Original Scale
Warning : BCa Intervals used Extreme Quantiles
Some BCa intervals may be unstable
```

The warning messages arise because in the extreme tails of the bootstrap distribution it is very difficult to make precise estimates—in such cases one really needs more bootstrap samples to obtain somewhat reliable estimates. Nevertheless, one can see that the intervals agree reasonably well; the BCα intervals are slightly shifted downward compared to the percentile intervals. For this coefficient, in absolute value the largest of the set, neither contains zero, as expected. In total, the percentile intervals show 318 cases where zero is not in the 95% confidence interval; the BCα intervals lead to 325 such cases.

It is interesting to repeat this exercise using a really large number of principal components, say twenty (remember, the gasoline data set only contains sixty samples). We would expect much more variation in the coefficients, since the model is more flexible and can adapt to changes in the training data much more easily. More variation means wider confidence intervals, and fewer "significant" cases, where zero is not included in the CI. Indeed, using twenty PCs leads to only 71 significant cases for the percentile intervals, and 115 for BCα (and an increased number of warning messages from the boot function as well).

9.6.3   Other R Packages for Bootstrapping

The bootstrap is such a versatile technique, that it has found application in many different areas of science. This has led to a large number of R packages implementing some form of the bootstrap—at the moment of writing, the package list of the CRAN repository contains already four other packages in between the packages **boot** and **bootstrap** already mentioned. To name just a couple of examples: package **FRB** contains functions for applying bootstrapping in robust statistics; **DAIM** provides functions for error estimation including the 0.632 and 0.632+ estimators. Using **EffectiveDose** it is possible to estimate the effects of a drug, and in particular to determine the effective dose level—bootstrapping is provided for the calculation of confidence intervals. Packages **meboot** and **BootPR** provide machinery for the application of bootstrapping in time series.

9.7 Integrated Modelling and Validation

Obtaining a good multivariate statistical model is hardly ever a matter of just loading the data and pushing a button: rather, it is a long and sometimes seemingly endless iteration of visualization, data treatment, modelling and validation. Since these aspects are so intertwined, it seems to make sense to develop methods that combine them in some way. In this section, we consider approaches that combine elements of model fitting with validation. The first case is bagging (Breiman 1996), where many models are fitted on bootstrap sets, and predictions are given by the average of the predictions of these models. At the same time, the out-of-bag samples can be used for obtaining an unbiased error estimate. Bagging is applicable to all classification and regression methods, but will give benefits only in certain cases; the classical example where it works well is given by trees (Breiman 1996)—see below. An extension of bagging, also applied to trees, is the technique of random forests (Breiman 2001). Finally, we will look at boosting (Freund and Schapire 1997), an iterative method for binary classification giving progressively more weight to misclassified samples. Bagging and boosting can be seen as meta-algorithms, because they consist of strategies that, in principle at least, can be combined with any model-fitting algorithm.

9.7.1 Bagging

The central idea behind bagging is simple: if you have a classifier (or a method for predicting continuous variables) that on average gives good predictions but has a somewhat high variability, it makes sense to average the predictions over a large number of applications of this classifier. The problem is how to do this in a sensible way: just repeating the same fit on the same data will not help. Breiman proposed to use bootstrapping to generate the variability that is needed. Training a classifier on every single bootstrap sets leads to an ensemble of models; combining the predictions of these models would then, in principle, be closer to the true answer. This combination of bootstrapping and aggregating is called bagging (Breiman 1996).

The package **ipred** implements bagging for classification, regression and survival analysis using trees—the **rpart** implementation is employed. For classification applications, also the combination of bagging with kNN is implemented (in function ipredknn). We will focus here on bagging trees. The basic function is ipredbag, while the function bagging provides the same functionality using a formula interface. Making a model for predicting the octane number for the gasoline data is very easy:

```
> (gasoline.bagging <- ipredbagg(gasoline$octane[gas.odd],
+                                 gasoline$NIR[gas.odd, ],
+                                 coob = TRUE))

Bagging regression trees with 25 bootstrap replications
Out-of-bag estimate of root mean squared error:   0.9181
```

The OOB error is quite high. Predictions for the even-numbered samples can be obtained by the usual `predict` function:

```
> gs.baggpreds <-
+    predict(gasoline.bagging, gasoline$NIR[gas.even, ])
> resids <- gs.baggpreds - gasoline$octane[gas.even]
> sqrt(mean(resids^2))
[1] 1.6738
```

This is not a very good result. Nevertheless, one should keep in mind that default settings are often suboptimal and some tweaking may lead to substantial improvements.

Doing classification with bagging is equally simple. Here, we show the example of discriminating between the `control` and `pca` classes of the prostate data, again using only the first 1000 variables as we did in Sect. 7.1.6.1:

```
> prost.bagging <- bagging(type ~ ., data = prost.df,
+                          subset = prost.odd)
> prost.baggingpred <- predict(prost.bagging,
+                              newdata = prost.df[prost.even, ])
> table(prost.type[prost.even], prost.baggingpred)
          prost.baggingpred
           control pca
   control      30  10
   pca           4  80
```

which doubles the number of misclassifications compared to the SVM solution in Sect. 7.4.1 but still is a lot better than the single-tree result.

So when does bagging improve things? Clearly, when a classification or regression procedure changes very little with different bootstrap samples, the result will be the same as the original predictions. It can be shown (Breiman 1996) that bagging is especially useful for predictors that are unstable, i.e., predictors that are highly adaptive to the composition of the data set. Examples are trees, neural networks (Hastie et al. 2001) or variable selection methods.

9.7.2 Random Forests

The combination of bagging and tree-based methods is a good one, as we saw in the last section. However, Breiman and Cutler saw that more improvement could be obtained by injecting extra variability into the procedure, and they proposed a number of modifications leading to the technique called Random Forests (Breiman 2001). Again, bootstrapping is used to generate data sets that are used to train an ensemble of trees. One key element is that the trees are constrained to be very simple—only few nodes are allowed, and no pruning is applied. Moreover, at every split, only a subset of all variables is considered for use. Both adaptations force diversity into the ensemble, which is the key to why improvements can be obtained with aggregating.

It can be shown (Breiman 2001) that an upper bound for the generalization error is given by

$$\hat{E} \le \bar{\rho}(1 - q^2)/q^2$$

where $\bar{\rho}$ is the average correlation between predictions of individual trees, and q is a measure of prediction quality. This means that the optimal gain is obtained when many good yet diverse classifiers are combined, something that is intuitively logical—there is not much point in averaging the outcomes of identical models, and combining truly bad models is unlikely to lead to good results either.

The R package **randomForest** provides a convenient interface to the original Fortran code of Breiman and Cutler. The basic function is `randomForest`, which either takes a formula or the usual combination of a data matrix and an outcome vector:

```
> wines.df <- data.frame(vint = vintages, wines)
> (wines.rf <- randomForest(vint ~ ., subset = wines.odd,
+                     data = wines.df))

Call:
 randomForest(formula = vint ~ ., data = wines.df, subset = wines.odd)
               Type of random forest: classification
                     Number of trees: 500
No. of variables tried at each split: 3

        OOB estimate of  error rate: 4.49%
Confusion matrix:
           Barbera Barolo Grignolino class.error
Barbera         24      0          0    0.000000
Barolo           0     28          1    0.034483
Grignolino       2      1         33    0.083333
```

The `print` method shows the result of the fit in terms of the error rate of the out-of-bag samples, in this case less than 5%. Because the algorithm fits trees to many different bootstrap samples, this error estimate comes for free. Prediction is done in the usual way:

```
> wines.rf.predict <-
+     predict(wines.rf, newdata = wines.df[wines.even, ])
> table(wines.rf.predict, vintages[wines.even])

wines.rf.predict Barbera Barolo Grignolino
       Barbera        24      0          0
       Barolo          0     29          0
       Grignolino      0      0         35
```

So prediction for the even rows in the data set is perfect here. Note that repeated training may lead to small differences because of the randomness involved in selecting bootstrap samples and variables in the training process. Also in many other applications random forests have shown very good predictive abilities (see, e.g., reference Svetnik et al. 2003 for an application in chemical modelling).

So it seems the most important disadvantage of tree-based methods, the generally low quality of the predictions, has been countered sufficiently. Does this come at a price? At first sight, yes. Not only does a random forest add complexity to the original algorithm in the form of tuning parameters, the interpretability suffers as well. Indeed, an ensemble of trees would seem more difficult to interpret than one simple sequence of yes/no questions. Yet in reality things are not so simple. The interpretability, one of the big advantages of trees, becomes less of an issue when one realizes that a slight change in the data may lead to a completely different tree, and therefore a completely different interpretation. Such a small change may, e.g., be formed by the difference between successive crossvalidation or bootstrap iterations—thus, the resulting error estimate may be formed by predictions from trees using different variables in completely different ways.

The technique of random forests addresses these issues in the following ways. A measure of the importance of a particular variable   is obtained by comparing the out-of-bag errors for the trees in the ensemble with the out-of-bag errors when the values for that variable are permuted randomly. Differences are averaged over all trees, and divided by the standard error. If one variable shows a big difference, this means that the variable, in general, is important for the classification: the scrambled values lead to models with decreased predictivity. This approach can be used for both classification (using, e.g., classification error rate as a measure) and regression (using a value like MSE). An alternative is to consider the total increase in node purity.

In package **randomForest** this is implemented in the following way. When setting the parameter `importance = TRUE` in the call to `randomForest`, the importances of all variables are calculated during the fit—these are available through the extractor function `importance`, and for visualization using the function `varImpPlot`:

```
> wines.rf <- randomForest(vint ~ ., data = wines.df,
+                          importance = TRUE)
> varImpPlot(wines.rf)
```

The result is shown in Fig. 9.8. The left plot shows the importance measured using the mean decrease in accuracy; the right plot using the mean decrease in node impurity, as measured by the Gini index. Although there are small differences, the overall picture is the same using both indices.

The second disadvantage, the large number of parameters to set in using tree-based models, is implicitly taken care of in the definition of the algorithm: by requiring all trees in the forest to be small and simple, no elaborate pruning schemes are necessary, and the degrees of freedom of the fitting algorithm have been cut back drastically.

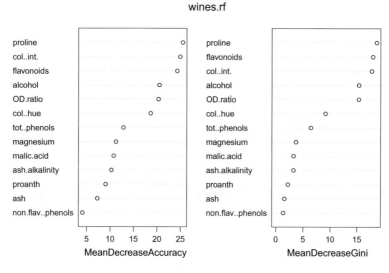

Fig. 9.8 Assessment of variable importance by random forests: the left plot shows the mean decrease in accuracy and the right the mean decrease in Gini index, both after permuting individual variable values

Furthermore, it appears that in practice random forests are very robust to changes in settings: averaging many trees also takes away a lot of the dependence on the exact value of parameters. In practice, the only parameter that is sometimes optimized is the number of trees (Efron and Hastie 2016), and even that usually has very little effect. This has caused random forests to be called one of the most powerful off-the-shelf classifiers available.

Just like the classification and regression trees seen in Sect. 7.3, random forests can also be used in a regression setting. Take the gasoline data, for instance: training a model using the default settings can be achieved with the following command.

```
> gasoline.rf <- randomForest(gasoline$NIR[gas.odd, ],
+                             gasoline$octane[gas.odd],
+                             importance = TRUE,
+                             xtest = gasoline$NIR[gas.even, ],
+                             ytest = gasoline$octane[gas.even])
```

For interpretation purposes, we have used the importance = TRUE argument, and we have provided the test samples at the same time. The results, shown in Fig. 9.9, are better than the ones from bagging:

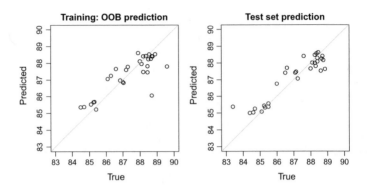

Fig. 9.9 Predictions for the gasoline data using random forests. Left plot: OOB predictions for the training data—right plot: test data

```
> pl.range <- c(83, 90)
> plot(gasoline$octane[gas.odd], gasoline.rf$predicted,
+       main = "Training: OOB prediction", xlab = "True",
+       ylab = "Predicted", xlim = pl.range, ylim = pl.range)
> abline(0, 1, col = "gray")
> plot(gasoline$octane[gas.even], gasoline.rf$test$predicted,
+       main = "Test set prediction", xlab = "True",
+       ylab = "Predicted", xlim = pl.range, ylim = pl.range)
> abline(0, 1, col = "gray")
```

However, there seems to be a bias towards the mean—the absolute values of the predictions at the extremes of the range are too small. Also the RMS values confirm that the test set predictions are much worse than the PLS and PCR estimates of 0.21:

```
> resids <- gasoline.rf$test$predicted - gasoline$octane[gas.even]
> sqrt(mean(resids^2))
[1] 0.63721
```

One of the reasons can be seen in the variable importance plot, shown in Fig. 9.10:

```
> rf.imps <- importance(gasoline.rf)
> plot(wavelengths, rf.imps[, 1] / max(rf.imps[, 1]),
+       type = "l", xlab = "Wavelength (nm)",
+       ylab = "Importance", col = "gray")
> lines(wavelengths, rf.imps[, 2] / max(rf.imps[, 2]), col = 2)
> legend("topright", legend = c("Error decrease", "Gini index"),
+        col = c("gray", "red"), lty = 1)
```

Both criteria are dominated by the wavelengths just above 1200 nm. Especially the Gini index leads to a sparse model, whereas the error-based importance values clearly are much more noisy. Interestingly, when applying random forests to the first derivative spectra of the gasoline data set (not shown) the same feature around 1200 nm is important, but the response at 1430 nm comes up as an additional feature.

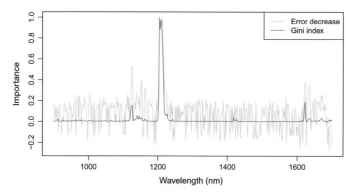

Fig. 9.10 Variable importance for modelling the gasoline data with random forests: basically, only the wavelengths just above 1200 nm seem to contribute

Although the predictions improve somewhat, they are still nowhere near the PLS and PCR results shown in Chap. 8.

For comparison, we also show the results of random forests on the prediction of the even samples in the prostate data set:

```
> prost.rf <-
+    randomForest(x = prost[prost.odd, ],
+                 y = prost.type[prost.odd],
+                 x.test = prost[prost.even, ],
+                 y.test = prost.type[prost.even])
> prost.rfpred <- predict(prost.rf, newdata = prost[prost.even, ])
> table(prost.type[prost.even], prost.rfpred)
         prost.rfpred
          control pca
  control      30  10
  pca           4  80
```

Again, a slight improvement over bagging can be seen.

9.7.3 Boosting

In boosting (Freund and Schapire 1997), validation and classification are combined in a different way. Boosting, and in particular in the adaBoost algorithm that we will be focusing on in this section, is an iterative algorithm that in each iteration focuses the attention to misclassified samples from the previous step. Just as in bagging, in principle any modelling approach can be used; also similar to bagging, not all combinations will show improvements. Other forms of boosting have appeared since the original adaBoost algorithm, such as gradient boosting, popular in the statistics community (Friedman 2001; Efron and Hastie 2016). One of the most powerful new variants is XGBoost (which stands for Extreme Gradient Boosting, Chen and Guestrin 2016), available in R through package **xgboost**.

The main idea of adaBoost is to use weights on the samples in the training set. Initially, these weights are all equal, but during the iterations the weights of incorrectly predicted samples increase. In adaBoost, which stands for adaptive boosting, the changes in the weight of object i is given by

$$D_{t+1}(i) = \frac{D_t(i)}{Z_t} \times \begin{cases} e^{-\alpha_t} & \text{if correct} \\ e^{\alpha_t} & \text{if incorrect} \end{cases} \tag{9.21}$$

where Z_t is a suitable normalization factor, and α_t is given by

$$\alpha_t = 0.5 \ln \left(\frac{1 - \epsilon_t}{\epsilon_t} \right) \tag{9.22}$$

with ϵ_t the error rate of the model at iteration t. In prediction, the final classification result is given by the weighted average of the T predictions during the iterations, with the weights given by the α values.

The algorithm itself is very simple and easily implemented. The only parameter that needs to be set in an application of boosting is the maximal number of iterations. A number that is too large would potentially lead to overfitting, although in many cases it has been observed that overfitting does not occur (see, e.g., references in Freund and Schapire 1997).

Boosting trees in R is available in package **ada** (Michailides et al. 2006), which directly follows the algorithms described in reference (Friedman et al. 2000). Let us revisit the prostate example, also tackled with SVMs (Sect. 7.4.1):

```
> prost.ada <- ada(type ~ ., data = prost.df, subset = prost.odd)
> prost.adapred <-
+    predict(prost.ada, newdata = prost.df[prost.even, ])
> table(prost.type[prost.even], prost.adapred)
          prost.adapred
           control pca
  control      30  10
  pca           3  81
```

The result is equal to the one obtained with bagging. The development of the errors in training and test sets can be visualized using the default `plot` command. In this case, we should add the test set to the `ada` object first[5]:

```
> prost.ada <- addtest(prost.ada,
+                       prost.df[prost.even, ],
+                       prost.type[prost.even])
> plot(prost.ada, test = TRUE)
```

---

[5]We could have added the test set data to the original call to `ada` as well—see the manual page.

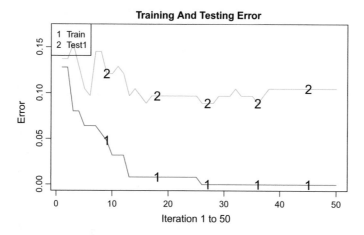

Fig. 9.11 Development of prediction errors for training and test sets of the prostate data (two classes, only 1000 variables) using `ada`

This leads to the plot in Fig. 9.11. The final error on the test set is less than half of the error at the beginning of the iterations. Clearly, both the training and testing errors have stabilized already after some twenty iterations.

The version of boosting employed in this example is also known as *Discrete adaboost* (Friedman et al. 2000; Hastie et al. 2001), since it returns 0/1 class predictions. Several other variants have been proposed, returning membership probabilities rather than crisp classifications and employing different loss functions. In many cases they outperform the original algorithm (Friedman et al. 2000).

Since boosting is in essence a binary classifier, special measures must be taken to apply it in a multi-class setting, similar to the possibilities mentioned in Sect. 7.4.1.1. A further interesting connection with SVMs can be made (Schapire et al. 1998): although boosting does not explicitly maximize margins, as SVMs do, it does come very close. The differences are, firstly, that SVMs use the L_2 norm, the sum of the squared vector elements, whereas boosting uses L_1 (the sum of the absolute values) and L_∞ (the largest value) norms for the weight and instance vectors, respectively. Secondly, boosting employs greedy search rather than kernels to address the problem of finding discriminating directions in high-dimensional space. The result is that although there are intimate connections, in many cases the models of boosting and SVMs can be quite different.

The obvious drawback of focusing more and more on misclassifications is that these may be misclassifications with a reason: outlying observations, or samples with wrong labels, may disturb the modelling to a large extent. Indeed, boosting has been proposed as a way to detect outliers.

Chapter 10
Variable Selection

Variable selection is an important topic in many types of multivariate modelling: the choice which variables to take into account to a large degree determines the result. This is true for every single technique discussed in this book, be it PCA, clustering methods, classification methods, or regression. In the unsupervised approaches, uninformative variables can obscure the "real" picture, and distances between objects can become meaningless. In the supervised cases (both classification and regression), there is the danger of chance correlations with dependent variables, leading to models with low predictive power. This danger is all the more real given the very low sample-to-variable ratios of many current data sets. The aim of variable selection then is to reduce the independent variables to those that contain relevant information, and thereby to improve statistical modelling. This should be seen both in terms of predictive performance (by decreasing the number of chance correlations) and in interpretability—often, models can tell us something about the system under study, and small sets of coefficients are usually easier to interpret than large sets.

In some cases, one is able to decrease the number of variables significantly by utilizing domain knowledge. A classical application is peak-picking in spectral data. In metabolomics, for instance, where biological fluids are analyzed by, e.g., NMR spectroscopy, one can typically quantify hundreds of metabolites. The number of metabolites is usually orders of magnitude smaller than the number of variables (ppm values) that have been measured; moreover, the metabolite concentrations lend themselves for immediate interpretation, which is not the case for the raw NMR spectra. A similar idea can be found in the field of proteomics, where mass spectrometry is used to find the presence or absence of proteins, based on the presence or absence of certain peptides. Quantification is more problematic here, so typically one obtains a list of proteins that have been found, including the number of fragments that have been used in the identification. When this step is possible it is nearly always good to do so. The only danger is to find what is already known—in many cases, data bases are used in the interpretation of the complex spectra: an unexpected compound, or a compound that is not in the data base but is present in the sample, is likely to be missed. Moreover, incorrect assignments present additional difficulties. Even so,

© Springer-Verlag GmbH Germany, part of Springer Nature 2020
R. Wehrens, *Chemometrics with R*, Use R!,
https://doi.org/10.1007/978-3-662-62027-4_10

the list of metabolites or proteins may be too long for reliable modelling or useful interpretation, and one is interested in further reduction of the data.

Very often, this variable selection is achieved by looking at the coefficients themselves: the large ones are retained, and variables with smaller coefficients are removed. The model is then refitted with the smaller set, and this process may continue until the desired number of variables has been reached. Unfortunately, as shown in Sect. 8.1.1, model coefficients can have a huge variance when correlation is high, a situation that is the rule rather than the exception in the natural sciences nowadays. As a result, coefficient size is not always a good indicator of importance. A more sophisticated approach is the one we have seen in Random Forests, where the decrease in model quality upon permutation of the values in one variable is taken as an importance measure. Especially for systems with not too many variables, however, tests for coefficient significance remain popular.

An alternative way of tackling variable selection is to use modelling techniques that explicitly force as many coefficients as possible to be zero: all these are apparently not important for the model and can be removed without changing the fitted values or the predictions. It can be shown that a ridge-regression type of approach with a penalty on the size of the coefficients has this effect, if the penalty is suitably chosen (Hastie et al. 2001)—a whole class of methods has descended from this principle, starting with the lasso (Tibshirani 1996).

One could say that the only reliable way of assessing the modelling power of a smaller set is to try it out—and if the result is disappointing, try out a different subset of variables. Given a suitable error estimate, one can employ optimization algorithms to find the subset that gives maximal modelling power. Two strategies can be followed: one is to fix the size of the subset, often dictated by practical considerations, and find the set that gives the best performance; the other is to impose some penalty on including extra variables and let the optimization algorithm determine the eventual size. In small problems it is possible, using clever algorithms, to find the globally optimal solution; in larger problems it very quickly becomes impossible to assess all possible solutions, and one is forced to accept that the global optimum may be missed.

10.1   Coefficient Significance

Testing whether coefficient sizes are significantly different from zero is especially useful in cases where the number of parameters is modest, less than fifty or so. Even if it does not always lead to the optimal subset, it can help to eliminate large numbers of variables that do not contribute to the predictive abilities of the model. Since this is a univariate approach—every variable is tested individually—the usual caveats about correlation apply. Rather than concentrating on the size and variability of individual coefficients, one can compare nested models with and without a particular variable. If the error decreases significantly upon inclusion of that variable, it can be said to be relevant. This is the basis of many stepwise approaches, especially in regression.

10.1.1 Confidence Intervals for Individual Coefficients

Let's use the wine data as an example, and predict class labels from the thirteen measured variables. We can assess the confidence intervals for the model quite easily, formulating the problem in a regression sense. For each of the three classes a regression vector is obtained. The coefficients for Grignolino, third class, can be obtained as follows:

```
> X <- wines[wines.odd, ]
> C <- classvec2classmat(vintages[wines.odd])
> wines.lm <- lm(C ~ X)
> wines.lm.summ <- summary(wines.lm)
> wines.lm.summ[[3]]
Call:
lm(formula = Grignolino ~ X)
Residuals:
    Min      1Q  Median      3Q     Max
-0.4657 -0.1387  0.0022  0.1326  0.4210

Coefficients:
                     Estimate Std. Error t value Pr(>|t|)
(Intercept)           2.77235    0.63633    4.36  4.1e-05 ***
Xalcohol             -0.12466    0.04918   -2.53   0.0133 *
Xmalic acid          -0.06631    0.02628   -2.52   0.0138 *
Xash                 -0.56351    0.12824   -4.39  3.6e-05 ***
Xash alkalinity       0.03227    0.00975    3.31   0.0014 **
Xmagnesium            0.00118    0.00173    0.68   0.4992
Xtot. phenols        -0.00434    0.07787   -0.06   0.9558
Xflavonoids           0.12497    0.05547    2.25   0.0272 *
Xnon-flav. phenols    0.36091    0.23337    1.55   0.1262
Xproanth              0.09320    0.05808    1.60   0.1128
Xcol. int.           -0.04748    0.01661   -2.86   0.0055 **
Xcol. hue             0.18276    0.16723    1.09   0.2779
XOD ratio             0.00589    0.06306    0.09   0.9258
Xproline             -0.00064    0.00012   -5.33  1.0e-06 ***
---
Signif. codes:  0 '***' 0.001 '**' 0.01 '*' 0.05 '.' 0.1 ' ' 1

Residual standard error: 0.209 on 75 degrees of freedom
Multiple R-squared:  0.847, Adjusted R-squared:  0.82
F-statistic: 31.9 on 13 and 75 DF,  p-value: <2e-16
```

The column with the stars in the output allows us to easily spot coefficients that are significant at a certain level. To get a summary of all variables that have p values smaller than, say, 0.1 for each of the three classes, we can issue:

```
> sapply(wines.lm.summ,
+        function(x) which(x$coefficients[, 4] < .1))
$'Response Barbera'
      Xmalic acid                    Xash          Xflavonoids
                3                       4                    8
Xnon-flav. phenols             Xcol. int.            XOD ratio
                9                      11                   13


$'Response Barolo'
      (Intercept)             Xalcohol       Xash Xash alkalinity
                1                    2          4               5
      Xflavonoids             Xproanth    XOD ratio        Xproline
                8                   10           13              14


$'Response Grignolino'
      (Intercept)             Xalcohol  Xmalic acid            Xash
                1                    2            3               4
   Xash alkalinity          Xflavonoids   Xcol. int.        Xproline
                5                    8           11              14
```

Variables ash and flavonoids occur as significant for all three cultivars; six others (not counting the intercept, of course) for two out of three cultivars.

In cases where no confidence intervals can be calculated analytically, such as in PCR or PLS, we can, e.g., use bootstrap confidence intervals. For the gasoline data, modelled with PCR using four latent variables, we have calculated bootstrap confidence intervals in Sect. 9.6.2. The percentile intervals, shown in Fig. 9.6, already indicated that most regression coefficients are significantly different from zero. How does that look for the (better) $BC\alpha$ confidence intervals? Let's find out:

```
> gas.BCACI <-
+   t(sapply(1:ncol(gasoline$NIR),
+            function(i, x) {
+                boot.ci(x, index = i, type = "bca")$bca[, 4:5]},
+            gas.pcr.bootCI))
```

A plot of the regression coefficients with these 95% confidence intervals (Fig. 10.1) immediately shows which variables are significantly different from zero:

```
> BCAcoef <- gas.pcr.bootCI$t0
> signif <- gas.BCACI[, 1] > 0 | gas.BCACI[, 2] < 0
> BCAcoef[!signif] <- NA


> matplot(wavelengths, gas.BCACI, type = "n",
+         xlab = "Wavelength (nm)",
+         ylab = "Regression coefficient",
+         main = "Gasoline data: PCR (4 PCs)")
> abline(h = 0, col = "gray")
> polygon(c(wavelengths, rev(wavelengths)),
+         c(gas.BCACI[, 1], rev(gas.BCACI[, 2])),
+         col = "pink", border = NA)
> lines(wavelengths, BCAcoef, lwd = 2)
```

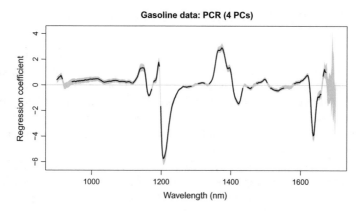

Fig. 10.1 Significance of regression coefficients for PCR using four PCs on the gasoline data; coefficients whose 95% confidence interval (calculated with the BCα bootstrap and indicated in pink) includes zero are not shown

Re-fitting the model after keeping only the 325 wavelengths leads to

```
> smallmod <- pcr(octane ~ NIR[, signif], data = gasoline,
+                 ncomp = 4, validation = "LOO")
> RMSEP(smallmod, intercept = FALSE, estimate = "CV")
1 comps  2 comps  3 comps  4 comps
 1.4342   1.4720   0.2756   0.2497
```

The error estimate is lower even than global minimum (at seven PCs) with the full data set containing 401 wavelengths. Here, one could also consider going for the three-component model which sacrifices very little in terms of RMSEP (it is still better than the seven-component model seen earlier) and has, well, one component fewer. After variable selection, refitting often leads to more parsimonious models in terms of the number of components needed. Even if the predictions are not (much) better, the improved interpretability is often seen as reason enough to consider variable selection.

Although this kind of procedure has been proposed in the literature several times, e.g., in Wehrens and van der Linden (1997), it is essentially incorrect. For the spectrum-like data, the correlations between the wavelengths are so large that the confidence intervals of individual coefficients are not particularly useful to determine which variables are significant—both errors of the first (false positives) and second kind (false negatives) are possible. Taking into account correlations and calculating so-called *Scheffé* intervals (Efron and Hastie 2016) often leads to intervals so wide that they have no practical relevance. The confidence intervals described above, for individual coefficients, at least give some idea of where important information is located.

10.1.2   Tests Based on Overall Error Contributions

In regression problems for data sets with not too many variables, the standard approach is stepwise variable selection. This can be performed in two directions: either one starts with a model containing all possible variables and iteratively discards variables that contribute least. This is called *backward selection*. The other option, *forward selection*, is to start with an "empty" model, i.e., prediction with the mean of the independent variable, and to keep on adding variables until the contribution is no longer significant.

As a criterion for inclusion, values like AIC, BIC or C_p can be employed—these take into account both the improvement in the fit as well as a penalty for having more variables in the model. The default for the R functions add1 and drop1 is to use the AIC. Let us consider the regression form of LDA for the wine data, leaving out the Barolo class for the moment:

```
> twowines.df <- data.frame(vintage = twovintages, twowines)
> twowines.lm0 <- lm(as.integer(vintage) ~ 1, data = twowines.df)
> add1(twowines.lm0, scope = names(twowines.df)[-1])
Single term additions

Model:
as.integer(vintage) ~ 1
                  Df Sum of Sq  RSS  AIC
<none>                         28.6 -168
alcohol            1    11.34 17.3 -226
malic.acid         1     8.75 19.9 -209
ash                1     3.15 25.5 -179
ash.alkalinity     1     1.07 27.6 -170
magnesium          1     0.72 27.9 -168
tot..phenols       1     7.57 21.1 -202
flavonoids         1    15.87 12.8 -262
non.flav..phenols  1     2.88 25.8 -178
proanth            1     4.69 23.9 -187
col..int.          1    18.07 10.6 -284
col..hue           1    15.27 13.4 -256
OD.ratio           1    17.94 10.7 -283
proline            1     3.70 24.9 -182
```

The dependent variable should be numeric, so in the first argument of the lm function, the formula, we convert the vintages to class numbers first. According to this model, the first variable to enter should be col..int—this gives the largest effect in AIC. Since we are comparing equal-sized models, this also implies that the residual sum-of-squares of the model with only an intercept and col..int is the smallest.

Conversely, when starting with the full model, the drop1 function would lead to elimination of the term that contributes least:

```
> twowines.lmfull <- lm(as.integer(vintage) ~ ., data = twowines.df)
> drop1(twowines.lmfull)
Single term deletions

Model:
as.integer(vintage) ~ alcohol + malic.acid + ash + ash.alkalinity +
    magnesium + tot..phenols + flavonoids + non.flav..phenols +
    proanth + col..int. + col..hue + OD.ratio + proline
                   Df Sum of Sq  RSS  AIC
<none>                           3.65 -387
alcohol            1    0.026   3.68 -388
malic.acid         1    0.331   3.98 -378
ash                1    0.127   3.78 -384
ash.alkalinity     1    0.015   3.67 -388
magnesium          1    0.000   3.65 -389
tot..phenols       1    0.098   3.75 -385
flavonoids         1    0.821   4.47 -364
non.flav..phenols  1    0.166   3.82 -383
proanth            1    0.028   3.68 -388
col..int.          1    0.960   4.61 -361
col..hue           1    0.162   3.81 -383
OD.ratio           1    0.254   3.91 -381
proline            1    0.005   3.66 -388
```

In this case, `magnesium` is the variable with the largest negative AIC value, and this is the first one to be removed.

Concentrating solely on forward or backward selection will in practice often lead to sub-optimal solutions: the order in which the variables are eliminated or included is of great importance and the chance of ending up in a local optimum is very real. Therefore, forward and backward steps are often alternated. This is the procedure implemented in the `step` function:

```
> step(twowines.lmfull, trace = 0)

Call:
lm(formula = as.integer(vintage) ~ malic.acid + ash + tot..phenols +
    flavonoids + non.flav..phenols + col..int. + col..hue + OD.ratio,
    data = twowines.df)

Coefficients:
      (Intercept)           malic.acid                  ash
           1.7220              -0.0571              -0.2359
      tot..phenols           flavonoids    non.flav..phenols
          -0.0833               0.2415               0.3821
         col..int.             col..hue             OD.ratio
          -0.0647               0.2236               0.1348
```

From the thirteen original variables, only eight remain.

Several other functions can be used for the same purpose: the **MASS** package contains functions `stepAIC`, `addterm` and `dropterm` which allows more model classes to be considered. Package **leaps** contains function `regsubsets`[1] which is guaranteed to find the best subset, based on the branch-and-bounds algorithm. Another package implementing this algorithm is **subselect**, with the function `eleaps`.

The branch-and-bounds algorithm was first proposed in 1960 in the area of linear programming (Land and Doig 1960), and was introduced in statistics by Furnival and Wilson (1974). The title of the latter paper has led to the name of the R-package. This particular algorithm manages to avoid many regions in the search space that can be shown to be less good than the current solution, and thus is able to tackle larger problems than would have been feasible using an exhaustive search. Application of the `regsubsets` function leads to the same set of selected variables (now we can provide a factor as the dependent variable):

```
> twowines.leaps <- regsubsets(vintage ~ ., data = twowines.df)
> twowines.leaps.sum <- summary(twowines.leaps)
> names(which(twowines.leaps.sum$which[8, ]))
[1] "(Intercept)"      "malic.acid"       "ash"
[4] "tot..phenols"     "flavonoids"       "non.flav..phenols"
[7] "col..int."        "col..hue"         "OD.ratio"
```

In some special cases, approximate distributions of model coefficients can be derived. For two-class linear discriminant analysis, a convenient test statistic is given by Mardia et al. (1979):

$$F = \frac{a_i^2(m - p + 1)c^2}{t_i m(m + c^2)D^2} \quad (10.1)$$

with $m = n_1 + n_2 - 2$, n_1 and n_2 signifying group sizes, p the number of variables, $c^2 = n_1 n_2/(n_1 + n_2)$, and D^2 is the Mahalanobis distance between the class centers, based on all variables. The estimated coefficient in the discriminant function is a_i, and t_i is the i-th diagonal element in the inverse of the total variance matrix T, given by

$$T = W + B \quad (10.2)$$

This statistic has an F-distribution with 1 and $m - p + 1$ degrees of freedom.

Let us see what that gives for the wine data without the Barolo samples. We can re-use the code in Sect. 7.1.3, now using all thirteen variables to calculate the elements for the test statistic:

---

[1]It also contains the function `leaps` for compatibility reasons; `regsubsets` is the preferred function.

```
> Tii <- solve(BSS + WSS)
> Ddist <- mahalanobis(colMeans(wines.groups[[1]]),
+                      colMeans(wines.groups[[2]]),
+                      wines.pcov12)
> m <- sum(sapply(wines.groups, nrow)) - 2
> p <- ncol(wines)
> c <- prod(sapply(wines.groups, nrow)) /
+   sum(sapply(wines.groups, nrow))
> Fcal <- (MLLDA^2 / diag(Tii)) *
+   (m - p + 1) * c^2 / (m * (m + c^2 * Ddist))
> which(Fcal > qf(.95, 1, m-p+1))
       malic.acid               ash          flavonoids
                2                 3                   7
non.flav..phenols          col..int.             col..hue
                8                10                  11
         OD.ratio
               12
```

Using this method, seven variables are shown to be contributing to the separation
between Grignolino and Barbera wines on the $\alpha = 0.05$ level. The only variable
missing, when compared to the earlier selected set of eight, is tot..phenols,
which has a p-value of 0.08.

10.2 Explicit Coefficient Penalization

In the chapter on multivariate regression we already saw that several methods use the
concept of shrinkage to reduce the variance of the regression coefficients, at the cost
of bias. Ridge regression achieves this by explicit coefficient penalization, as shown
in Eq. 8.22. Although it forces the coefficients to be closer to zero, the values almost
never will be exactly zero. If that would be the case, the method would be performing
variable selection: those variables with zero values for the regression coefficients can
safely be removed from the data.

Interestingly enough, one can obtain the desired behavior by replacing the
quadratic penalty in Eq. 8.22 by an absolute-value penalty:

$$\underset{B}{\operatorname{argmax}} (Y - XB)^2 + \lambda|B| \qquad (10.3)$$

The penalty, consisting of the sum of the absolute values of the regression coefficients,
is an L_1-norm. As already stated before, ridge regression, focusing on squared coefficients, employs an L_2-norm, and measures like AIC or BIC are using the L_0-norm,
taking into account only the number of non-zero regression coefficients. In Eq. 10.3,
with increasing values for parameter λ more and more regression coefficients will
be exactly zero. This method has become known under the name *lasso* (Tibshirani 1996; Hastie et al. 2001); an efficient method to solve this equation—and
related approaches—has become known under the name of *least-angle regression*,

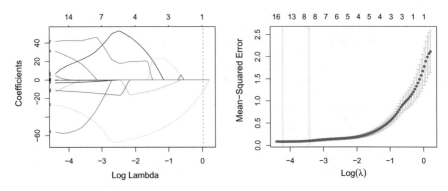

Fig. 10.2 Left: lasso model coefficients plotted against the (relative) penalty size λ. Right: validation plot

or LARS (Efron et al. 2004). Several R versions for the lasso are available. Package **glmnet** is written by the inventors of the method, and will be used here as an example. Other packages implementing similar techniques include **lars**, where slightly different defaults have been chosen for solving the lasso problem, **lpc** for "lassoed principal components" and **relaxo**, a generalization of the lasso using possibly different penalization coefficients for the variable selection and parameter estimation steps.

Rather than one set of coefficients for one given value of λ, the function `glmnet` returns an entire sequence of fits, with corresponding regression coefficients. For the odd rows of the gasoline data, the model is simply obtained as follows:

```
> gas.lasso <- glmnet(x = gasoline$NIR[gas.odd, ],
+                     y = gasoline$octane[gas.odd])
> plot(gas.lasso, xvar = "lambda", label = TRUE)
```

The result of the corresponding `plot` method is shown in the left panel of Fig. 10.2. It shows the (standardized) regression coefficients against the size of the L_1 norm of the coefficient vector. For an infinitely large value of λ, the weight of the penalty, no variables are selected. Gradually decreasing the penalty leads to a fit using only one non-zero coefficient. Its size varies linearly with the penalty—until the next variable enters the fray. The right of the plot shows the position of the entrances of new non-zero coefficients. This piecewise linear behavior is the key to the lasso algorithm, and makes it possible to calculate the whole trace in approximately the same amount of time as needed for a normal linear regression. Around the left axis (and somewhat hard to read in this default set-up), the variable numbers of some of the coefficients are shown at their "final" values, i.e., at the last value for λ, by default one percent of the value at which the first variable enters the model.

Of course, the value of the regularization parameter λ needs to be optimized. A function `cv.glmnet` is available for that, using by default ten-fold crossvalidation. Two common measures are available as predefined choices. Obviously, the model corresponding to the lowest crossvalidation error is one of them; the other is the most sparse model that is within one standard deviation from the global optimum (Hastie et al. 2001), the same criterion also used in the **pls** package for determining the optimal number of latent variables mentioned in Sect. 8.2.2.

```
> gas.lasso.cv <- cv.glmnet(gasoline$NIR[gas.odd, ],
+                           gasoline$octane[gas.odd])
> svals <- gas.lasso.cv[c("lambda.1se", "lambda.min")]
```

The plot command for the `cv.glmnet` object leads to the validation plot in the right panel of Fig. 10.2. The global minimum in the CV curve lies at a value of -4.215, and the one-se criterion at -3.424 (both in log units, as in the figure). The associated errors can be obtained directly using the `predict` function for the crossvalidation object:

```
> gas.lasso.preds <-
+    lapply(svals,
+            function(x)
+                predict(gas.lasso,
+                         newx = gasoline$NIR[gas.even, ],
+                         s = x))
> sapply(gas.lasso.preds,
+         function(x) rms(x, gasoline$octane[gas.even]))
lambda.1se lambda.min
   0.18881    0.19463
```

The prediction error for the test set using the optimal penalty is better than the best values seen with PCR and PLS, the one with the more conservative estimate somewhat larger. In both cases, only a very small subset of the original variables are included in the model:

```
> gas.lasso.coefs <- lapply(svals,
+                           function(x) coef(gas.lasso, s = x))
> sapply(gas.lasso.coefs,
+         function(x) sum(x != 0))
lambda.1se lambda.min
         9         14
```

A further development is mixing the L_1-norm of the lasso and related methods with the L_2-norm used in ridge regression. This is known as the *elastic net* (Zou and Hastie 2005). The penalty term is given by

$$\sum_i \left(\alpha |\beta_i| + (1 - \alpha)\beta_i^2 \right) \tag{10.4}$$

where the sum is over all variables. The result is that large coefficients are penalized heavily (because of the quadratic term) and that many of the coefficients are exactly zero, leading to a sparse solution.

The `glmnet` function provides ridge regression through specifying `alpha = 0` and the lasso with `alpha = 1`. It will be no surprise that values of `alpha` between zero and one lead to the elastic net:

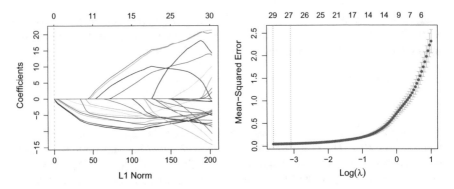

Fig. 10.3 Elastic net results for the gasoline data using $\alpha = 0.5$. The left plot shows the development of the regression coefficients upon relaxation of the penalty parameter. The right plot shows the ten-fold crossvalidation curve, optimizing λ

```
> gas.elnet <- glmnet(gasoline$NIR, gasoline$octane, alpha = .5)
> plot(gas.elnet, "norm")
```

The result is shown in the left plot in Fig. 10.3. Further inspection of the elastic net model, including the crossvalidation plot on the right side of Fig. 10.3, is completely analogous to the code shown earlier for the lasso. The performance of the elastic net in predicting the test set is slightly better than the lasso, at the expense of including more variables:

```
> sapply(gas.elnet.preds,
+         function(x) rms(x, gasoline$octane[gas.even]))
lambda.1se lambda.min
   0.15881    0.15285
> sapply(gas.elnet.coefs,
+         function(x) sum(x != 0))
[1] 28 30
```

The coefficients that are selected by the global-minimum lasso and elastic-net models are shown in Fig. 10.4. There is good agreement between the two sets; the elastic net in general selects variables in the same region as the lasso, with the exception of the area around 1000 nm with is not covered by the lasso at all. Note that the coefficient sizes for the elastic net are much smaller (in absolute size) than the ones from the lasso, a result of the L_2 penalization.

10.3   Global Optimization Methods

Given the speed of modern-day computing, it is possible to examine large numbers of different models and select the best one. However, as we already saw with leaps-and-bounds approaches, even in cases with a moderate number of variables it is practically

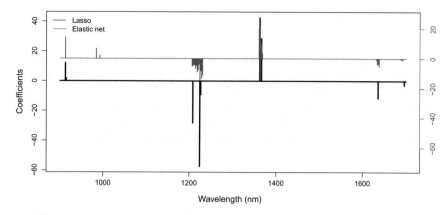

Fig. 10.4 Non-zero coefficients in the lasso and elastic net models. A small vertical offset has been added to facilitate the comparison

impossible to assess the quality of all subsets. One must, therefore, limit the number of subsets that is going to be considered to a manageable size. The stepwise approach does this by performing a very local search around the current best solution before adding or removing one variable; it can be compared to a steepest-descent strategy. The obvious disadvantage is that many areas of the search space will never be visited. For regression or classification cases with many variables, almost surely the method will find a local optimum, very often of low quality.

An alternative is given by random search—just sampling randomly from all possible subsets until time is up. Of course, the chance of finding the global optimum in this way is smaller than the chance of winning the lottery... What is needed is a search strategy that combines random elements with "gradient" information; that is, a strategy that uses information, available in solutions of higher quality, with the ability to throw that information away if needed, in order to be able to escape from local optima. This type of approaches has become known under the heading of *global search* strategies. The two best-known ones in the area of chemometrics are *Simulated Annealing* and *Genetic Algorithms*. Both will be treated briefly below.

What is quality, in this respect, again depends on the application. In most cases, the property of interest will be the quality of prediction of unseen data, which for larger data sets can conveniently be estimated by crossvalidation approaches. For data sets with few samples, this will not work very well because of the coarse granularity of the criterion: many subsets will lead to an equal number of errors. Additional information should be used to distinguish between these.

10.3.1 Simulated Annealing

In Simulated Annealing (SA, Kirkpatrick et al. 1983; Cerny 1985), a sequence of candidate solutions is assessed, starting from a random initial point. A new solution with quality E_{t+1}, not too far away from the current one (with quality E_t), is unconditionally accepted if it is better than the current one. If $E_t > E_{t+1}$ on the other hand, accepting the move corresponds to a deterioration. However, and this is the defining feature of SA, such a move can be accepted, with a probability equal to

$$p_{acc} = \exp\left(\frac{E_{t+1} - E_t}{T_t}\right) \tag{10.5}$$

where T_t the state of the control parameter at the current time point t. Note that p_{acc}, defined in this way, is always between zero and one (since $E_t > E_{t+1}$). This criterion is known as the *Metropolis* criterion (Metropolis et al. 1953). Other criteria are possible, too, but are rarely used.

The name Simulated Annealing comes from an analogy to annealing in metallurgy, where crystals with fewer defects can be created by repeatedly heating and cooling a material: during the (slow) cooling, the atoms are able to find their energetically most favorable positions in a regular crystal lattice, whereas the heating allows atoms that have been caught in unfavorable positions (local optima) to "try again" in the next cooling stage. The analogy with the optimization task is clear: if an improvement is found (better atom positions) it is accepted; if not, then sometimes a deterioration in quality is accepted in order to be able to cross a ridge in the solution landscape and to find an solution that is better in the end. Very often, the control parameter is therefore indicated with T, to stress the analogy with temperature. During the optimization, it will slowly be decreasing in magnitude—the cooling—causing fewer and fewer solutions of lower quality to be accepted. In the end, only real improvements are allowed. It can be shown that SA leads to the global optimum if the cooling is slow enough (Granville et al. 1994); unfortunately, the practical importance of this proof is limited since the cooling may have to be infinitely slow. Note that random search is a special case that can be achieved simply by setting T_t to an extremely large value, leading to $p_{acc} = 1$ whatever the values of E_{t+1} and E_t.

The naive implementation of an SA therefore can be very simple: one needs a function that generates a new solution in the neighborhood of the current one, an evaluation function to assess the quality of the new solution, and the acceptance function, including a cooling schedule for the search parameter T. The evaluation needs to be defined specifically for each problem. In regression or classification cases typically some estimate of prediction accuracy is used such as crossvalidation—note that the evaluation function in this schedule probably is the most time-consuming step, and since it will be executed many times (typically thousands or, in complicated cases, even millions of solutions are evaluated by global search methods) it should be very fast. If enough data are available then one could think of using a separate test set for the evaluation, or of using quality criteria such as Mallows's C_p, or AIC or BIC values, mentioned in Chap. 9. The whole SA algorithm can therefore easily be summarized in a couple of steps:

1. Choose a starting temperature and state;
2. Generate and evaluate a new state;
3. Decide whether to accept the new state;
4. Decrease the temperature parameter;
5. Terminate or go to step 2.

Several SA implementations are available in R. We will have a look at the `optim` function from the core **stats** package which implements a general-purpose SA function.

Let us see how this works in the two-class wines example from Sect. 10.1.2, excluding the Barolo variety. This is a simple example for which it still is quite difficult to assess all possible solutions, especially since we do not force a model with a specific number of variables. We will start with the general-purpose `optim` approach, since this provides most insight in the inner workings of the SA. First we need to define an evaluation function. Here, we use the fast built-in LOO classification estimates of the `lda` function:

```
> lda.loofun <- function(selection, xmat, grouping, ...) {
+    if (sum(selection) == 0) return(100)
+    lda.obj <- lda(xmat[, selection == 1], grouping, CV = TRUE)
+    100*sum(lda.obj$class != grouping)/length(grouping)
+ }
```

Argument `selection` is a vector of numbers here, with ones at the position of the selected variables, and zeroes elsewhere. Since `optim` by default does minimization, the evaluation function returns the percentage of misclassified cases—note that if no variables are selected, a value of 100 is returned.

Now that we have defined what exactly we are going to optimize, we need to define a step function, leading from the current solution to the next. A simple approach could be to do one of three things: either remove a variable, add a variable, or replace a variable. If too few variables are selected, we could increase the number by adding one previously unselected variable randomly (so the escape clause in the evaluation function checking for zero selected variables should never be reached). That seems easy enough to put in a function:

```
> saStepFun <- function(selected, ...) {
+    maxval <- length(selected)
+    selection <- which(selected == 1)
+    newvar2 <- sample(1:maxval, 2)
+
+    ## too short: add a random number
+    if (length(selection) < 2) {
+      result <- unique(c(selection, newvar2))[1:2]
+    } else {       # generate two variable numbers
+      presentp <- newvar2 %in% selection
+      ## if both are in x, remove the first
+      if (all(presentp)) {
+        result <- selection[selection != newvar2[1]]
+      } else {    # if none are in selection, add the first
+        if (all(!presentp)) {
+          result <- c(selection, newvar2[1])
+        } else { # otherwise swap
+          result <- c(selection[selection != newvar2[presentp]],
+                      newvar2[!presentp])
+        }}}
+
+    newselected <- rep(0, length(selected))
+    newselected[result] <- 1
+    newselected
+ }
```

Both in the evaluation and step function we use the ellipses (. . .) to prevent undefined
arguments to throw errors: optim simply transfers all arguments that are not its own
to both underlying functions, where they can be used or ignored.

We will start with a random subset of five columns. This leads to the following
misclassification rate:

```
> initselect <- rep(0, ncol(wines))
> initselect[sample(1:ncol(wines), 5)] <- 1
> (r0 <- lda.loofun(initselect, x = twowines,
+                   grouping = twovintages))
[1] 2.521
```

This corresponds to 3 misclassifications. How much can we improve using simulated
annealing? Let's find out:

```
> SAoptimWines <-
+   optim(initselect,
+         fn = lda.loofun, gr = saStepFun, method = "SANN",
+         x = twowines, grouping = twovintages)
```

The result is a simple list with the first two elements containing the best result and
the corresponding evaluation value:

```
> SAoptimWines[c("par", "value")]
$par
 [1] 1 0 0 0 1 0 1 1 0 1 1 0 1

$value
 [1] 0
```

In this case, all misclassifications have been eliminated while still using only a subset of the variables, in this case 7 columns. We could try to push the number of selected variables back by adding a small penalty for every selected variable—once the ideal value of zero misclassifications has been reached the current definition of the evaluation function gives no more opportunities for further improvement.

By default, 10,000 evaluations are performed in the optim version of SA; this number can be changed using the control argument, where also the initial temperature and the cooling rate can be adjusted. In real, nontrivial problems, it will probably take some experimentation to find optimal values for these search parameters.

A more ambitious example is to predict the octane number of the gasoline samples with only a subset of the NIR wavelengths. The step function is the same as in the wine example, and the only thing we have to do is to define the evaluation function:

```
> pls.cvfun <- function(selection, xmat, response, ncomp, ...) {
+    if (sum(selection) < ncomp) return(Inf)
+    pls.obj <- plsr(response ~ xmat[, selection == 1],
+                    validation = "CV", ncomp = ncomp, ...)
+    c(RMSEP(pls.obj, estimate = "CV", ncomp = ncomp,
+           intercept = FALSE)$val)
+ }
```

In this case, we use the explicit crossvalidation provided by the plsr function. This adds a little bit of variability in the evaluation function since repeated application will lead to different segments—but the savings in time are quite big. We will assume that this variability is smaller than the gains we hope to make. The number of components to take into account can be specified in the extra argument of the evaluation function; the error of the model with the largest number of latent variables is returned. The RMSEP function returns an object of class mvrVal, where the val list element contains the numerical value of interest—this is what we will return. Now, let us try to find an optimal two-component PLS model (fewer variables often lead to less complicated models). We start with a very small model using only eight variables $(.02 \times 401)$:

```
> nNIR <- ncol(gasoline$NIR)
> initselect <- rep(0, nNIR)
> initselect[sample(1:nNIR, 8)] <- 1
> SAoptimNIR1 <-
+    optim(initselect,
+          fn = pls.cvfun, gr = saStepFun, method = "SANN",
+          x = gasoline$NIR, response = gasoline$octane,
+          ncomp = 2, maxval = nNIR)
```

```
> pls.cvfun(initselect, gasoline$NIR, gasoline$octane, ncomp = 2)
[1] 1.3896
> (nvarSA1 <- sum(SAoptimNIR1$par))
[1] 190
> SAoptimNIR1$value
[1] 0.24823
```

The result is already quite good: compare this, e.g., to the values in the left panel in
Fig. 8.4 (where we were looking at the crossvalidation of a model based on the odd
samples only) and it is clear that the estimated error with fewer variables and two
components is less than half that of the two-component model including all variables.

Still, we see that the number of variables included in the model is quite high—
perhaps more sparse models can be found that are equally good or even better. In
such cases, it pays to abandon the naive approach adopted above and look closer at
the problem itself. We should realize we are optimizing a long parameter vector in
this case, with 401 values. Many of these values are zero to start with, and we would
like to retain the sparsity of the solution. Our step function, however, is not taking this
into account and will suggest many steps leading to more variables. Combine that
with a rather high initial temperature parameter, and it is clear that especially in the
beginning many bad moves will be accepted. Finally, the evaluation function does
not reward sparse solutions explicitly. Let's see what a lower starting temperature
and an adapted evaluation function contribute. First we will define the latter:

```
> pls.cvfun2 <- function(selection, xmat, response, ncomp,
+                        penalty = 0.01, ...) {
+   if (sum(selection) < ncomp) return(Inf)
+   pls.obj <- plsr(response ~ xmat[, selection == 1, drop = FALSE],
+                   validation = "CV", ncomp = ncomp, ...)
+   c(RMSEP(pls.obj, estimate = "CV", ncomp = ncomp,
+          intercept = FALSE)$val) +
+     penalty * sum(selection)
+ }
```

Note that we need at least ncomp variables in order to fit a PLS model. Next, we
will define a step function that, using the default settings, will keep the number of
variables approximately equal to the starting situation. By playing with the cutoffs
in the argument plimits (a random draw from the uniform distribution lower than
the first value will lead to eliminating one of the selected variables; anything larger
than the second number to the addition of a variable, and anything in between to
swapping variables) one can tweak the behavior of the step function:

```
> saStepFun2 <- function(selected, plimits = c(.3, .7), ...) {
+   dowhat <- runif(1)
+
+   ## decrease selection
+   if (dowhat < plimits[1]) {
+     if (sum(selected) > 2) { # not too small...
+       kickone <- sample(which(selected == 1), 1)
+       selected[kickone] <- 0
+       return(selected)
+     }
+   }
+
+   ## increase selection
+   if (dowhat > plimits[2]) { # not too big...
+     if (sum(selected) < length(selected)) {
+       addone <- sample(which(selected == 0), 1)
+       selected[addone] <- 1
+       return(selected)
+     }
+   }
+
+   ## swap
+   kickone <- sample(which(selected == 1), 1)
+   selected[kickone] <- 0
+   addone <- sample(which(selected == 0), 1)
+   selected[addone] <- 1
+   selected
+ }
```

By changing the values of the `plimits` argument we can directly influence the number of nonzero entries in the result: e.g., the higher the first number, the bigger the chance that a variable will be removed. Let's see how that works, combined with a lower starting temperature. The default is 10—we will try a value of 1:

```
> penalty <- 0.01
> SAoptimNIR2 <-
+   optim(initselect,
+         fn = pls.cvfun2, gr = saStepFun2, method = "SANN",
+         x = gasoline$NIR, response = gasoline$octane,
+         ncomp = 2, maxval = nNIR,
+         control = list(temp = 1))
```

This leads to the following results:

```
> (nvarSA2 <- sum(SAoptimNIR2$par))
[1] 6
> SAoptimNIR2$value - penalty*nvarSA2
[1] 0.20047
```

Now we have a crossvalidation error that is clearly better than what we saw earlier but with only 6 instead of 190 variables in the model.

Several packages provide SA functions specifically optimized for variable selection. The `anneal` function in package **subselect**, e.g., can be used for variable selection in situations like discriminant analysis, PCA, and linear regression, according to the criterion employed. For LDA, this function takes the between-groups covariance matrix, the minimal and maximal number of variables to be selected, the within-groups covariance matrix and its expected rank, and a criterion to be optimized (see below) as arguments. For the wine example above, a solution to find the optimal three-variable subset would look like this:

```
> winesHmat <- ldaHmat(twowines.df[, -1], twowines.df[, 1])
> wines.anneal <-
+     anneal(winesHmat$mat, kmin = 3, kmax = 3,
+             H = winesHmat$H, criterion = "ccr12", r = 1)


> wines.anneal$bestsets
        Var.1 Var.2 Var.3
Card.3      2     7    10
> wines.anneal$bestvalues
  Card.3
0.83281
```

Repeated application (using, e.g., `nsol = 10`) in this case leads to the same solution every time. Rather than the direct estimates of prediction error, the `anneal` function uses functions of the within- and between-groups covariance matrices (Silva 2001). In this case using the `ccr12` criterion, the first root of BW^{-1} is optimized, analogous to Fisher's formulation of LDA in Sect. 7.1.3. As an other example, Wilk's Λ is given by

$$\Lambda = \det(W)/\det(T) \tag{10.6}$$

and is (in a slightly modified form) available in the `tau2` criterion. For the current case where the dimensionality of the within-covariance matrices is estimated to be one, all criteria lead to the same result.

The new result differs from the subset from our own implementation in only one instance: variable 11, color hue, is swapped for the malic acid concentration. The reason, of course, is that both functions optimize different criteria. Let us see how the two solutions fare when evaluated with the criterion of the other algorithm. The value for the `ccr12` criterion of the solution using variables 7, 10 and 11, found with our own simplistic SA implementation, can be assessed easily:

```
> ccr12.coef((nrow(twowines.df) - 1) * var(twowines.df[, -1]),
+             winesHmat$H, r = 1, c(7, 10, 11))
[1] 0.82293
```

which, as expected, is slightly lower than that of the set consisting of variables 2, 7 and 10. Conversely, the prediction quality of the newer set is slightly worse (two misclassifications):

```
> selection <- rep(0, ncol(twowines))
> selection[c(2, 7, 10)] <- 1
> lda.loofun(selection, twowines.df[, -1], twowines.df[, 1])
[1] 1.6807
```

Obviously, there are probably many sets with the same or similar values for the quality criterion of interest, and to some extent it is a matter of chance which one is returned by the search algorithm. Moreover, the number of possible quality values can be limited, especially with criteria based on the number of misclassifications. This can make it more difficult to discriminate between two candidate subsets.

The `anneal` function for subset selection is also applicable in other types of problems than classification alone: e.g., for variable selection in PCA it uses a measure of similarity of the original data matrix and of the projections on the k-variable subspace—again, several different criteria are available. The speed and applicability in several domains are definite advantages of this particular implementation. However, there are some disadvantages, too: firstly, because of the formulation using covariance matrices it is hard to apply `anneal` to problems with large numbers of variables. Finding the most important discriminating variables in the prostate data set would stretch your computer to the limit—in fact, even the gasoline example requires the argument `force = TRUE` since the default is to refuse cooperation (and give a serious-looking warning) as soon as the number of variables exceeds 400.

Secondly, the function does not allow one to submit an evaluation function, and one has to do with the predefined set—crossvalidation-based approaches such as used in the examples above cannot be implemented, increasing the danger of overfitting. Finally, it can be important to monitor the progress of the optimization, or at least keep track of the speed with which improvements are found—especially when fine-tuning the SA parameters (temperature, cooling rate) one would like to have the possibility to assess acceptance rates. Currently, no such functionality is provided in the **subselect** package.

One other dedicated SA approach for variable selection can be found in the **caret** package mentioned in Chap. 7 in the form of the `safs` (simulated annealing feature selection) function. This function does allow crossvalidation-based quality measures to guide the optimization, but also supports external test sets and criteria like AIC. Parallelization is supported at several different levels.

10.3.2 Genetic Algorithms

Genetic Algorithms (GAs, Goldberg 1989) manage a population of candidate solutions, rather than one single solution as is the case with most other optimization methods. Every solution in the population is represented as a string of values, and in a process called cross-over, mimicking sexual reproduction, offspring is generated combining parts of the parent solutions. Random mutations, occurring with relatively low frequency, ensure that some diversity is maintained in the population. The quality of the offspring is measured in an evaluation phase—again in analogy with biology, this quality is often called "fitness". Strings with a low fitness will have no or only a low probability of reproduction, so that subsequent generations will generally consist of better and better solutions. This obvious imitation of the process of natural selection has led to the name of the technique. GAs have been applied to a wide range of problems in very diverse fields—several overviews of applications within chemistry can be found in the literature (e.g., Leardi 2001; Niazi and Leardi 2012).

Just like with Simulated Annealing, GAs need an evaluation function to obtain fitness values for trial solutions. A step function, on the other hand, is not needed: the genetic machinery (cross-over and mutation operations) will take care of that. Several parameters need to be set, such as the size of the population, the number of iterations, and the chances of crossover and mutation, but that is all. Population sizes are typically in the order of 50–100; the number of iterations in the order of several hundreds. There are some aspects, however, that are particular for GAs. The first choice we have to make is on the *representation* of the candidate solutions, i.e., the candidate subsets. For variable selection, two obvious possibilities present themselves: either a vector of indices of the variables in the subset, or a string of zeros and ones. For other optimization problems, e.g., non-linear fitting, real numbers can also be used. Secondly, the *selection* function needs to be defined. This determines which solutions are allowed to reproduce, and is the driving force behind the optimization—if all solutions would have the same probability the result would be a random search. Typical selection procedures are to use random sampling with equal probabilities for all solutions above a quality cutoff, or to use random sampling with (scaled) quality indicators as probability weights.

The **GA** package (Scrucca 2013, 2017) provides a convenient and efficient toolbox, supporting for binary, real-valued and permutation representations, and several standard genetic operators. In addition, users can define their own operators. Parallel evaluation of population members is supported (especially useful if the evaluation of a single solution takes some time), and to speed up proceedings even further, local searches can be allowed at random intervals to inject new and useful information in the population. Finally, populations can be "seeded", i.e., one can provide one or more solutions that are thought to be approximately correct.

Applying the ga function from the **GA** package to our gasoline data is quite easy. We can use the same evaluation function as used in the SA optimization, pls.cvfun2, where a small penalty is applied for solutions with more variables. Since ga does maximization only, we multiply the result with −1:

```
> fitnessfun <- function(...) -pls.cvfun2(...)
```

Now we are ready to go. The simplest approach would be to apply standard procedures and hope for the best:

```
> GAoptimNIR1 <-
+   ga(type = "binary", fitness = fitnessfun,
+      x = gasoline$NIR, response = gasoline$octane,
+      ncomp = 2, penalty = penalty,
+      nBits = ncol(gasoline$NIR), monitor = FALSE, maxiter = 100)
```

The result, as we may have expected, still contains many variables, and has a high crossvalidation error:

```
> (nvarGA1 <- sum(GAoptimNIR1@solution))
[1] 149
> -GAoptimNIR1@fitnessValue + penalty*nvarGA1
[1] 3.2732
```

Ouch, that does not look too good. Of course we have not been fair: the random initialization of the GA will lead to a population with approximately 50% selected variables, where the initial SA solution had only 2%. In addition, the default mutation function is biased to this 50% ratio as well: in sparse solutions it is much more likely to add a variable than to remove one. Similar to the adaptation of the step function in SA, we define the mutation function in such a way that setting bits to zero is (much) more likely than setting bits to one, a behavior that can be controlled by the value of the bias argument:

```
> myMutate <- function (object, parent, bias = 0.01)
+ {
+     mutate <- parent <- as.vector(object@population[parent, ])
+     n <- length(parent)
+     probs <- abs(mutate - bias)
+     j <- sample(1:n, size = 1, prob = probs)
+     mutate[j] <- abs(mutate[j] - 1)
+     mutate
+ }
```

In the **GA** package these settings are controlled by the gaControl function, and changes remain in effect for the rest of the session (or until changed again). Including the new mutation function and using a more reasonable initial state is easily done:

```
> gaControl("binary" = list(mutation = "myMutate"))
> popSize <- 50 # default
> initmat <- matrix(0, popSize, nNIR)
> initmat[sample(1:(popSize*nNIR), nNIR)] <- 1
>
> GAoptimNIR2 <-
+   ga(type = "binary", fitness = fitnessfun,
+       x = gasoline$NIR, response = gasoline$octane,
+       popSize = popSize, nBits = ncol(gasoline$NIR),
+       ncomp = 2, suggestions = initmat, penalty = penalty,
+       monitor = FALSE, maxiter = 100)
```

This leads to the following result:

```
> (nvarGA2 <- sum(GAoptimNIR2@solution))
[1] 4
> -GAoptimNIR2@fitnessValue + penalty*nvarGA2
[1] 0.26728
```

Clearly, this constitutes a substantial improvement over the first optimization result, getting close to the SA solution presented earlier. Of course, more experimentation can easily lead to further improvements (as is the case with SA as well).

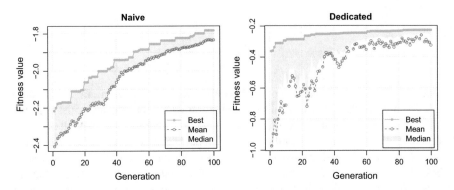

Fig. 10.5 GA optimization results for the NIR data. Left panel: naive application; right panel: application with a specific initialization matrix and a dedicated mutation function. Note the different *y* scales

One of the nice features of the **GA** package is that the results of calling ga can be plotted as well. Figure 10.5 shows the optimization trajectories of both GA runs. First, note the difference in the *y*-axes: the dedicated GA leads to much better fitnesses. The plots show the best result (green dots/line) as well as the average and the mean fitness values of the population at each iteration. If the latter two are very close to the best value, there is too little variation in the population and the result is likely to be quite bad. Especially with the dedicated mutation operator, one sees quite sudden jumps when "worse" solutions are introduced in the population (too few variables even lead to Inf values in which case no Mean is displayed), but still these solutions may contain kernels of good information.

The curves also give an idea on whether it is useful to put in additional effort: the left panel of Fig. 10.5 clearly gives the impression that further improvements are possible. In most cases, playing around with search parameters or tweaking the fitness function will have more chance of reaching good results than simply increasing the number of iterations.

In more complicated problems, speed is a big issue. Some simple tricks can be employed to speed up the optimization. Typically, several candidate solutions will survive unchanged during a couple of generations. Rigorous bookkeeping may prevent unnecessary quality assessments, which in almost all cases is the most computer-intensive part of a GA. An implementational trick that is also very often applied is to let the best few solutions enter the next generation unchanged; this process, called *elitism*, makes sure that no valuable information is thrown away and takes away the need to keep track of the best solution. Provisions can be taken to prevent premature convergence: if the population is too homogeneous the power of the crossover operator decreases drastically, and the optimization usually will not lead to a useful answer. One strategy is to disallow offspring that is equal to other candidate solutions; a second strategy is to penalize the fitness of candidate solutions that are too similar; the latter strategy is sometimes called *sharing*.

Other GA implementations are available, too, of course. The **caret** package includes a `gafs` function that is very similar to the `safs` function we saw earlier for SA. The `genetic` function in the **subselect** package provides a fast Fortran-based GA. The details of the crossover and mutation functions are slightly different from the description above—indeed, there are probably very few implementations that share the exact same crossover and mutation operators, testimony to the flexibility and power of the evolutionary paradigm. Having seen the working of the `anneal` function, most input parameters will speak for themselves:

```
> wines.genetic <-
+    genetic(winesHmat$mat, kmin = 3, kmax = 5, nger = 20,
+            popsize = 50, maxclone = 0,
+            H = winesHmat$H, criterion = "ccr12", r = 1)
> wines.genetic$bestvalues
 Card.3  Card.4  Card.5
0.83281 0.84368 0.85248
> wines.genetic$bestsets
        Var.1 Var.2 Var.3 Var.4 Var.5
Card.3      2     7    10     0     0
Card.4      2     3     7    10     0
Card.5      2     3     7    10    12
```

And indeed, the same three-variable solution is found as the optimal one. This time, also four- and five-variable solutions are returned (because of the values of the `kmin` and `kmax` arguments).

The `maxclone` argument tries to enforce diversity by replacing duplicate off-spring by random solutions (which are not checked for duplicity, however). Leaving out this argument would, in this simple example, lead to a premature end of the optimization because of the complete homogeneity of the population. Both `anneal` and `genetic` provide the possibility of a further local optimization of the final best solution.

10.3.3 Discussion

Variable selection is a difficult process. Simple stepwise methods only work with a small number of variables, whereas the largest gains can be made in the nowa-days typical situation of hundreds or even thousands of variables. More complicated methods containing elements of random search, such as SA or GA approaches, can have a high variability, especially in cases where correlations between variables are high. One approach is to repeat the variable selection multiple times, and to use those variables that are consistently selected. Although this strategy is intuitively appeal-ing, it does have one flaw: suppose that variables a and b are highly correlated, and that a combination of either a or b with a third variable c leads to a good model. In repeated selection runs, c will typically be selected twice as often as a or b—if the overall selection threshold is chosen to include c but neither of a and b, the model will not work well.

In addition, the optimization criterion is important. It has been shown that LOO crossvalidation as a criterion for variable selection is inconsistent, in the sense that even with an infinitely large data set it will not choose the correct model (Shao 2003). Baumann et al. advocate the use of leave-multiple-out crossvalidation for this purpose (Baumann et al. 2002a, b), even though the computational burden is high. In this approach, the data are repeatedly split, randomly, in training and test sets, where the number of repetitions needs to be greater than the number of variables, and for every split a separate crossvalidation is performed to optimize the parameters of the modelling method such as the number of latent variables in PCR or PLS. A workable alternative is to fix the number of latent variables to a "reasonable" number, and to find the subset of variables that with this particular setting leads to the best results. This takes away the nested crossvalidation but may lead to subsets that are suboptimal. In general, one should accept the fact that there is no guarantee that the optimal subset will be found, and it is wise to accept a subset that is "good enough".

Part V
Applications

Chapter 11
Chemometric Applications

This chapter highlights some typical examples of research themes in the chemometrics community. Up to now we have concentrated on fairly general techniques, found in many textbooks and applicable in a wide range of fields. The topics in this chapter are more specific to the field of chemometrics, combining elements from the previous chapters. In particular, latent-variable approaches like PCA and PLS exhibit a wide range of applications (some people have criticized the field of chemometrics of being too preoccupied with latent-variable methods, and not without reason—on the other hand such tools are extremely handy in many different situations).

To start with, we come back to the problem of *missing values*. Hard to avoid in many real-life applications, they often prevent the standard application of statistical methods. One example is PCA—the svd-based implementation in Chap. 4 does not allow missing values. We will discuss a couple of alternatives, e.g., replacing the missing values with estimated values. Ironically, PCA is one of the methods that can be used to obtain these estimates... Another form of PCA, *robust PCA*, is an attractive method to identify outliers in multivariate space, at modest computational cost. It is very often a good idea to check whether a robust alternative (if it exists) leads to results that are close to what one sees in the main analysis: if that is not the case, one should really try to find the cause(s) of the differences and then decide what to do. Robust estimates also play a role in the next topic, statistical process control, which is very important in industrial applications. Again, multivariate approaches based on distances or dimension reduction firmly place this topic in the chemometrics area. Continuing the theme of finding ways to combat flaws in our data, *Orthogonal Signal Correction* and its combination with PLS, *OPLS*, provide ways to remove irrelevant variation in the data—irrelevant for predicting purposes, that is. In some cases this leads to simpler models that are easier to interpret. In analytical laboratories, there is often a need to develop calibration models that can be transferred across a range of instruments. One example is to develop a model using a laboratory, high-quality setup, and then to apply the model for in-line measurements of a much lower quality. The approach to achieve this has become known as *calibration transfer*. Finally, we

© Springer-Verlag GmbH Germany, part of Springer Nature 2020   249
R. Wehrens, *Chemometrics with R*, Use R!,
https://doi.org/10.1007/978-3-662-62027-4_11

take a look at a decomposition of a matrix X where the individual components are directly interpretable, e.g., as concentration profiles or spectra of pure compounds: *Multivariate Curve Resolution.*

11.1   PCA in the Presence of Missing Values

Real-life data sets nearly always contain missing values, i.e., data points for which no value has been recorded. Data analyses often cannot handle these missing values, and the regular approach is to replace the missing values with some hopefully appropriate estimate, and do the analysis on the completed data. This process is usually referred to as *imputation*, and is often repeated many times (multiple imputation, using different imputed values) to decrease the influence of the artificial values. In analytical chemical applications, a common cause for missing values is the detection or quantification limit of the measurement device: concentrations may simply be too low to lead to a measurable response. In other cases, values may be missing because of non-responses in surveys, errors in data processing (e.g., misalignments in LC-MS data), temporary breakdown of a sensor, or simply because of some random event—there are countless possible reasons why a data point is missing.

That does not mean it is not important to think about reasons for missingness; in fact, ignoring this is dangerous and can easily lead to false conclusions. Missing values being caused by measurements below the detection limit form a good example. We know that in these cases the true but unknown value should be somewhere between zero and the limit of detection. That is, even the missing values contain information. The simplest possible approach in such a case would be to pick a value somewhere in the middle and use that instead of all missing values. In many applications, such an approach is too simple, and leads to an awkward peak around the imputed value in the distribution of the variable. A better strategy is to try to estimate—on the basis of the non-missing values for a particular variable—the parameters of the distribution of a variable, and then draw randomly from that distribution to complete the data set (Uh et al. 2008). Again, this can be done multiple times, allowing to assess the effect of the imputed values on the analysis. An overview of many different ways of imputing data can be found, e.g., in Little and Rubin (2019).

In case there is reason to believe that the data are *missing completely at random* (a term so important in the field that the acronym MCAR is often used) life becomes simpler. MCAR means that there is no relation between the values of the data and the missingness status. This is clearly not the case for the detection limit example. Also the term *missing at random* (MAR) is used, and although this again implies that there is no relation between the values and the missingness status, it is different from MCAR in that missingness may depend on non-missing values. As a hypothetical example, hospital lab tests may show more missing values for obese patients than for patients with a lower BMI. This obviously makes it important to take into account these dependencies when imputing. The final category is *missing not at random* (MNAR), e.g., corresponding to the measurements below the detection limit mentioned earlier. In such a situation there is no easy way out and the mechanism leading to missingness needs to be taken into account in the analysis.

Many methods are available to handle missing values in a general context (Little and Rubin 2019). To name just two: the analysis may be based on only the complete cases, which may work well when the number of missing values is limited. It does run the risk of strongly biased results. Alternatively, missing values may be replaced by adequate values such as means, or estimated using methods like regression or the EM algorithm mentioned in the context of model-based clustering, described in Sect. 6.3—there, the unknown class label is basically treated as a missing value. Several R packages are available for more general situations. The mice package (van Buuren and Groothuis-Oudshoorn 2011), for example, assumes that data are MAR, and uses regressions to estimate missing values from the other variables. The name stands for Multivariate Imputation via Chained Equations. Categorical as well as numerical values are allowed. Another package for multiple imputation, amelia (named after Amelia Earhart, the first woman to fly across the Atlantic Ocean solo who went missing over the Pacific in 1937), also assumes MAR data but in addition assumes multivariate normal data (Honaker et al. 2011).

In the remainder of this paragraph we will focus on a couple of ways to perform PCA in the presence of missing data. First of all, we could sacrifice some of the functionality of PCA and use simple tricks allowing us to use the incomplete matrix anyway. Second, we could replace the missing values by something that makes sense, in the case of MCAR data perhaps a mean value. In all cases, it is probably wise to eliminate rows or columns that contain too many missing values – finding an optimal cutoff here is a trial-and-error process which will depend strongly on the application.

Let's look at the arabidopsis data from **ChemometricsWithR**, an LC-MS-based metabolomics data set on a number of samples of *Arabidopsis thaliana*, a popular model organism in plant sciences. As usual in this kind of data, many values are missing—the total number per variable is shown in Fig. 11.1. In the following, we will only retain variables with less than 40% missing values:

```
> data(arabidopsis)
> naLimitPerc <- 40
> naLimit <- floor(nrow(arabidopsis) * naLimitPerc / 100)
> nNA <- apply(arabidopsis, 2, function(x) sum(is.na(x)))
> naIdx <- which(nNA < naLimit)
> X.ara <- arabidopsis[, naIdx]
```

This leads to a matrix containing 249 columns, less than half of the number of original variables. We expect a large majority of missing values to be caused by metabolites being too low in concentration to be measured.

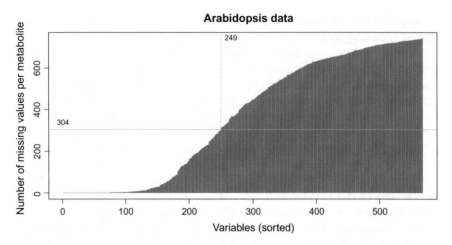

Fig. 11.1 Number of missing values in the LC-MS `arabidopsis` data (sorted). The cutoff of including a variable has been set here at no more than 40% missing values (i.e., at most 304 NAs)

11.1.1 Ignoring the Missing Values

The first approach is simply to ignore the missing values. This can be done by calculating the covariance or correlation matrix with the argument `use = "pairwise.complete.obs"`. Note that this (incorrectly!) assumes that the missing data are MCAR: one pretends that the correlations or covariances calculated with the subset of points that *is* observed on average does not differ from the values that would be calculated from the full matrix. As long as there are enough cases for which pairwise data are available, this will lead to a square matrix without any NA values from which scores or loadings can be derived, as explained in Sect. 4.2. A similar result would be obtained by calculating distances using only pairwise complete observations and then doing PCA on the distance matrix (PCoA, see Sect. 4.6.1).

Let's see how this works out for the arabidopsis data. First we need to decide on the scaling. Since the intensities are basically counts, we use log scaling, more or less the default in MS-based metabolomics data. Then there are several other choices that could be relevant here, such as Pareto scaling. For the sake of demonstration we will proceed with the most common choice which is autoscaling, giving each variable equal weight in the PCA:

```
> X.ara.l <- log(X.ara)
> X.ara.l.sc <- scale(X.ara.l)
```

Next we calculate the covariance matrix (which is equal to the correlation matrix here, because of the scaling we applied), check that the number of NA values is zero, and run `svd`:

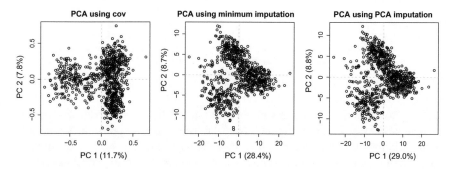

Fig. 11.2 PCA scoreplots from the arabidopsis data set using three different ways of handling missing data

```
> X.ara.cov <- cov(t(X.ara.l.sc), use = "pairwise.complete.obs")
> sum(is.na(X.ara.cov))
[1] 0
> X.ara.svd <- svd(X.ara.cov, nu = 2, nv = 2)
> ara.PCA.svd <-
+    structure(
+      list(scores = X.ara.svd$u %*% diag(sqrt(X.ara.svd$d[1:2])),
+           var = X.ara.svd$d,
+           totalvar = sum(X.ara.svd$d),
+           centered.data = TRUE),
+      class = "PCA")
```

Since the goal here is visualization, we limit the number of singular vectors to be calculated to two. The result is stored as an object of class PCA so that we can use the scoreplot.PCA function, leading to the left panel in Fig. 11.2:

```
> scoreplot(ara.PCA.svd, main = "PCA using cov")
```

The two-component model leads to a reasonable amount of variance explained in the first two components, given that the data matrix has 249 columns. Some structure seems to be visible, especially along the first axis.

11.1.2   Single Imputation

We already discussed that the method in the previous section assumes an MCAR regime which is unlikely for the current situation. Rather, we expect most of the values to be missing because of low concentrations. Replacing NA values with the smallest number in the column would therefore seem a more sensible idea:

```
> X.ara.imput1 <-
+   apply(X.ara, 2,
+         function(x) {
+             x[is.na(x)] <- min(x, na.rm = TRUE)
+             x
+         })
```

It is easy to check that the number of NAs after imputation is zero. Now that we have completed our matrix we can proceed using the standard PCA approach. Note that we perform autoscaling only *after* having done the imputation: obviously, the imputed values will affect column means and standard deviations. The resulting scoreplot is shown in the middle panel of Fig. 11.2:

```
> ara.PCA.minimputation <- PCA(scale(log(X.ara.imput1)))
```

We see the same structure with two or three clusters as in the previous case, but now rotated by something like 45°. The percentage of variance explained by the first PC in particular is much higher than in the case ignoring the missing values altogether.

Interestingly, a more elaborate version of single imputation can be done by PCA methods. One would start with imputing random values, perform a PCA, and reconstructing the values at the locations of the NAs with the values predicted by PCA. This process iterates until some convergence threshold is met. In this way, correlation structure is taken into account. Note that the number of PCs is again a parameter that needs to be set: in this case, models with two PCs are no longer subsets of models with more PCs, so one has to explicitly calculate the results for all dimensionalities. One function implementing this is imputePCA from the **missMDA** package (Josse and Husson 2016). Let's see how things go when we use two dimensions:

```
> X.ara.pcaimput <- imputePCA(X.ara.1, ncp = 2)$completeObs
```

The PCA scoreplot based on the PCA imputation is shown in the right panel of Fig. 11.2:

```
> ara.PCA.pcaimputation <- PCA(scale(X.ara.pcaimput))
```

It is very similar to the one obtained by imputing the missing values with column minima.

In the PCA-based imputation we have used log-scaled data, under the assumption that the data after log-transformation are perhaps a little bit more regular—in the other case that did not matter since we were taking the smallest value for each column. A very natural question is: what are the imputed values? Figure 11.3 shows the histograms. It is clear that the PCA-imputed values cover a much wider range than the column minima. Still, they are at the lower end of the data range.

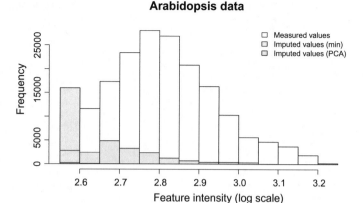

Fig. 11.3 Distribution of measured and imputed values in the `arabidopsis` data set

11.1.3 Multiple Imputation

Obviously we would like to assess how much our imputed values influence the scores. One way of doing this is to impute multiple times, and plot the scores as a much bigger point cloud. For the imputations, a more elaborate mechanism is needed than simply taking the mean or smallest value per column (repeating such an action would not be very informative). Function `MIPCA` from package **missMDA** (Josse and Husson 2016) provides several possible strategies. The default is to start with the imputed matrix from the iterative PCA algorithm in the previous paragraph. Parametric bootstrapping is used to sample residuals (assuming a normal distribution around zero, with the standard deviation given by the empirical standard deviation of the PCA residuals) which are used to generate (by default one hundred) bootstrap samples. Each of these bootstrap samples is then subjected to PCA. Finally, so-called Procrustes analysis is used to rotate all PCA solutions in such a way that the data are maximally overlapping (Josse et al. 2011).

To see how the method is applied we concentrate on the first twenty columns, where the following columns contain missing values:

```
> rownames(X.ara.1) <- rep("", nrow(X.ara.1))
> colnames(X.ara.1) <- paste("V", 1:ncol(X.ara.1), sep = "")
> countNAs <- apply(X.ara.1[, 1:20], 2, function(x) sum(is.na(x)))
> countNAs[countNAs > 0]
 V6  V8 V10 V13 V15 V16 V17 V19 V20
220 259 260 223 139 252  16 246   3
```

We would expect most variability in the variables containing many missing values. Let's apply `MIPCA`:

```
> ara.PCA.Minput <- MIPCA(X.ara.1[, 1:20], ncp = 2, scale = TRUE)
```

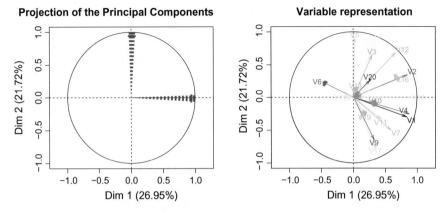

Fig. 11.4 Multiple imputation for the first twenty columns in the log-scaled `arabidopsis` data. The plot on the left shows that in each bootstrap sample the PCs are defined in a very similar way. The loading plot on the right shows point clouds for variables containing missing values—again the effect of the imputation seems limited here

The result is an object of class MIPCA which allows a variety of plots. Here we show two of them in Fig. 11.4, created by:

```
> plot(ara.PCA.Minput, choice = "dim", new.plot = FALSE)
> plot(ara.PCA.Minput, choice = "var", new.plot = FALSE)
```

The first one shows the variability in the definition of the principal components in the individual bootstrap sets. Clearly, there is little variation: both PC1 and PC2 remain very much in the same position. The second plot, a loading plot, confirms this: for those variables containing missing values a small point cloud can be seen around the loading arrow indicating the variability of the estimate due to bootstrap sets. For variables without missing values, no variability is observed.

11.2   Outlier Detection with Robust PCA

Identifying outliers, i.e., samples that do not conform to the general structure of the data, is a difficult and dangerous task, prone to subjective judgements. Once one has detected an outlier, or even several outliers, the question is what to do: should one remove these before further analysis, downweight their importance in some way, or simply leave them as they are and pay special attention to things like residuals? All are sensible strategies, and could be valid choices depending on the question at hand and the data available. Generally one is advised to not remove outliers, unless there are very good reasons to do so. Very often these reasons are required to include meta-information: deviating numbers in themselves may not be enough reason to remove outlying observations, but if one also knows that there was a power cut in the lab just before that particular measurement, or something else happened that may be related to this sample, then it may be more easy to decide to not take this particular record into account.

At the same time, it is important to realize that outlying samples *will* occur in practice, also if everything seems to have gone according to plan in the lab: whole microarrays with expressions of tens of thousands of genes can be useless because of some experimental artifact, and including them could be detrimental to the results. One of the problems is that if several outliers are present, they may make each other seem "normal", an effect that is called *masking*. Additionally, high-dimensional space, as we know, is mostly empty and every object of a small-to-medium-sized data set can be seen as an outlier. Only if we can assume that the samples are occupying a restricted subspace we may have hopes of performing meaningful outlier detection.

The area of *robust statistics* (Maronna et al. 2005) is a rich and flourishing field in which methods are studied that are less affected by individual values and will yield consistent results even in the presence of a sizeable fraction of outliers. A typical example of a robust location estimator is the median. Its value will not change if all data points above the median are suddenly ten units higher, or multiplied by a factor of one thousand. It is said to have a breakdown point of 0.5, meaning that half of the data can be "wrong" without affecting the estimate. Higher breakdown points than 0.5 obviously do not make too much sense. Note that the average as an estimator of location has a breakdown point of 0.0: any change to the measurements will lead to a different result. Many classical estimators have robust counterparts, that typically rely on fewer assumptions. The price to pay is usually a lower accuracy or a loss of power: typically one would need more samples to obtain comparable results. Robust methods therefore decrease the influence of outlying observations – interestingly, this makes them also very suited to identify these observations in the first place.

11.2.1   Robust PCA

From the above it may seem natural to conclude that a robust form of PCA would be a good candidate to identify outliers in a multivariate data set. In some cases even classical PCA will work: what would be easier than to apply PCA to the data, and see the outliers far away from the bulk of the data? Although this sometimes does happen, and PCA in these cases is a valuable outlier detection method, in other cases the outliers are harder to spot. The point is that PCA is not a robust method: since it is based on the concept of variance, outliers will greatly influence scores and loadings, sometimes even to the extent that they will dominate the first PCs. What is needed in such cases is a *robust* form of PCA (Hubert 2009). Many different approaches exist, each characterized by their own breakdown point, the fraction of outliers that can be present without influencing the covariance estimates.

The simplest form is to perform the SVD on a robust estimate of the covariance or correlation matrix (Croux and Haesbroeck 2000). One such estimate is given by the Minimum Covariance Determinant (MCD, Rousseeuw 1984), which has a breakdown point of up to 0.5. As the name already implies, the MCD estimator basically samples subsets of the data of a specific size, in search of the subset that leads to a covariance matrix with a minimal determinant, i.e., covering the smallest hypervolume. The assumption is that the outlying observations are far away from the other data points, increasing the volume of the covariance ellipsoid. The size of the subset, to be chosen by the user, determines the breakdown point, given by $(n - h + 1)/n$, with n the number of observations and h the size of the subset. Unless one really expects a large fraction of the data to be contaminated, it is recommended to choose $h \approx 0.75n$. The resampling approach can take a lot of time, and although fast algorithms are available (Rousseeuw and van Driessen 1999), matrices with more than a couple of hundred variables remain hard to tackle.

The MCD covariance estimator is available in several R packages. One example is `cov.mcd` in package **MASS**. If we use this in combination with the `princomp` function, we can see the difference between robust and classical covariance estimation. Let's focus on the Grignolino samples from the wine data:

```
> X <- wines[vintages == "Grignolino", ]
> X.sc <- scale(X)
> X.clPCA <- princomp(X.sc)
> X.robPCA <- princomp(X.sc, covmat = cov.mcd(X.sc))
```

Visualization using biplots leads to Fig. 11.5:

```
> biplot(X.clPCA, main = "Classical PCA")
> biplot(X.robPCA, main = "MCD-based PCA")
```

There are clear differences in the first two PCs: in the classical case PC 1 is dominated by the variables OD ratio, flavonoids, proanth and tot. phenols, leading to samples 63, 66, 15 and 1, 2, and 3 to having extreme coordinates. In the robust version, on the other hand, these samples have very relatively small PC 1 scores. Rather, they are extremes of the second component, the result of increased influence of variables (inversely) correlated with ash on the first component. Although many of the relations in the plots are similar (the main effect seems to be a rotation), the example shows that even in cases where one would not expect it applying (more) robust methods can lead to appreciable differences.

An important impediment for the application of the MCD estimator is that it can only be calculated for non-fat data matrices, i.e., matrices where the number of samples is larger than the number of variables—in other cases, the covariance matrix is singuar, with a determinant of zero. In such cases another approach is necessary. One example is ROBPCA (Hubert et al. 2005), combining Projection Pursuit and robust covariance estimation: PP is employed to find a subspace of lower dimension in which the MCD estimator can be applied. ROBPCA has one property that we also saw in ICA (Sect. 4.6.2): if we increase the number of PCs there is no guarantee that the first PCs will remain the same—in fact, they usually are not. Obviously, this can make interpretation somewhat difficult, especially since the method to choose the "correct" number of PCs is less obvious in robust PCA than in classical PCA (Hubert 2009).

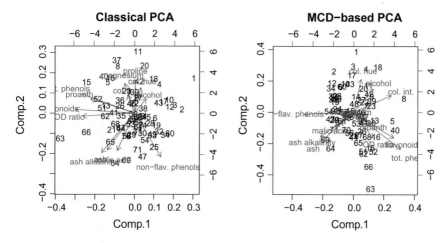

Fig. 11.5 Biplots for the Grignolino samples: the classical PCA solution is shown on the left, whereas the right plot is based on the MCD covariance estimate

Since the details of the ROBPCA algorithm are a lot more complicated than can be treated here, we just illustrate its use. ROBPCA, as well as several other robust versions of PCA, is available in package **rrcov** as the function PcaHubert. Application to the Grignolino samples using five PCs leads to the following result:

```
> X.HubPCA5 <- PcaHubert(X.sc, k = 5)
> summary(X.HubPCA5)

Call:
PcaHubert(x = X.sc, k = 5)
Importance of components:
                       PC1   PC2   PC3   PC4   PC5
Standard deviation   1.766 1.502 1.217 1.039 0.923
Proportion of Variance 0.355 0.257 0.169 0.123 0.097
Cumulative Proportion  0.355 0.612 0.780 0.903 1.000
```

Note that the final line gives the cumulative proportion of variance as a fraction of the variance captured in the robust PCA model, and not as the fraction of the total variance, usual in classical PCA. If we do not provide an explicit number of components (the default, k = 0) the algorithm chooses the optimal number itself:

```
> X.HubPCA <- PcaHubert(X.sc)
> summary(X.HubPCA)

Call:
PcaHubert(x = X.sc)
Importance of components:
                       PC1   PC2   PC3   PC4   PC5    PC6    PC7
Standard deviation   1.751 1.537 1.276 1.125 0.9862 0.8695 0.6062
Proportion of Variance 0.294 0.227 0.156 0.121 0.0934 0.0726 0.0353
Cumulative Proportion  0.294 0.521 0.677 0.799 0.8922 0.9647 1.0000
```

Fig. 11.6 Outlier map for the Grignolino data, based on a seven-component ROBPCA model

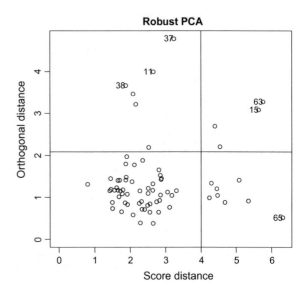

Apparently this optimal number equals seven in this case. The rule-of-thumb to calculate the "optimal" number of components is based on the desire to explain a significant portion of the variance explained by the model (a fraction of 0.8 is used as the default) while not taking into account components with very small standard deviations—the last component of the model should have an eigenvalue at least .1% of the largest one. If the number of variables is small enough, the MCD algorithm is used directly; if not, the ROBPCA algorithm is used. One can force the use of ROBPCA by setting mcd = FALSE. Note that the standard deviations of the first components are not the same as the ones calculated for the five-component model.

The default plotting method is different from the classical plot: it shows an outlier map, or distance-distance map, rather than scores or loadings. The main idea of this plot is to characterise every sample by two different distances:

- the Orthogonal Distance (OD), indicating the distance between the true position of every data point and its projection in the space of the first few PCs;
- the Score Distance (SD), or the distance of the sample projection to the center of all sample projections.

Or, put differently, the SD of a sample is the distance to the center, measured in the hyperplane of the PCA projection, and the OD is the distance to this hyperplane. Obviously, both SD and OD depend on the number of PCs. When a sample is above the horizontal threshold it is too far away from the PCA subspace; when it is to the right from the vertical threshold it is too far from the other samples *within* the PCA subspace. The horizontal and vertical thresholds are derived from χ^2 approximations (Todorov and Filzmoser 2009). For the Grignolino data, this leads to the plot in Fig. 11.6:

```
> plot(X.HubPCA)
```

Several of the most outlying samples are indicated with their indices, so that they can be inspected further. Also a biplot method is available, which shows a plot that is very similar to the right plot in Fig. 11.5. Inspection of the data shows that objects 63 and 15 do contain some extreme values in some of the variables—indeed, object 63 is also the object with the smallest score on PC 1 in a classical PCA. However, it would probably be too much to remove them from the data completely.

11.2.2 Discussion

A robust approach can be extremely important in cases where one suspects that some of the data are outliers. Classical estimates can be very sensitive to extreme values, and it frequently occurs that only one or very few samples dominate the rest of the data. This need not be an error, because influential observations may be correct, but in general one would put more trust in a model that is based on many observations rather than a few. This is not in contradiction with the desire to build sparse models, as seen in the section on SVMs, for example: there, the sparseness was obtained by selecting only those objects in the relevant part of the space, using all other objects in the selection process.

The robust methods in this section have a wider applicability than just outlier detection: they can be used as robust plugin estimators in classification and regression methods. Robust LDA can be obtained, for example, by using a robust estimate of the pooled covariance matrix; robust QDA by using robust covariances for all classes. PCR can be robustified in several ways, e.g., by applying SVD to a robust covariance matrix estimate; an alternative is formed by regressing on robust scores, for instance from the ROBPCA algorithm. One can even replace the least squares regression by robust regression methods such as least trimmed squares (Rousseeuw 1984). Also robust versions of PLS regression exist (Hubert and Branden 2003; Liebmann et al. 2009). These robust versions of classification and regression methods share the big advantage that one can safely leave in all objects, even though some of them may be suspected outliers: the analysis will not be influenced by only a couple atypical observations. And to turn the question of outliers around: if robust and classical analyses give the same or similar results, then one can conclude that there are no (influential) outliers in the data.

Note that here we have concentrated on identifying whole records as outlying observations, i.e., rows in our data matrix. This is not the only way to approach the issue. One could also say that certain variables, columns in the data matrix, show deviating behaviour. This is a situation, however, that is less likely to wreak havoc: many multivariate methods, especially in supervised approaches, are geared towards obtaining the optimal weights for each of the variables. If the outlying column would lead to worse results it would probably get a low weight anyway. In unsupervised approaches such as PCA the variable would stand out, and one then can easily identify it as a potential problem and decide how to tackle it. Only in distance- or kernel-based methods we would run the risk of obtaining suboptimal results. Finally, it is also possible—in fact, rather likely – that individual data values are grossly incorrect, for whatever reason. Since these have less influence on the overall model than outlying whole rows, in many cases they can be disregarded. However, approaches have been developed recently identifying such cases (Rousseeuw and den Bossche 2018).

R contains many packages with facilities for robust statistics, the most important one probably being **robustbase**. According to the taskview on CRAN, plans exist to further streamline the available packages, using **robustbase** as the basic package for robust statistics, and several more specialized packages building on that, such as is the case already for packages like **rrcov**.

11.3 Multivariate Process Monitoring

Robust methods like the PCA methods from the previous paragraph try to focus on the big picture, simply ignoring individual data points that do no conform to the general trend. However, such data points may also contain valuable information, especially when occuring in groups or in a particular order: then, they may point to imminent changes in process conditions that are not always beneficial. The idea is that these deviations, when noticed early enough, can be corrected for by changing appropriate process parameters. In this simple way, a control mechanism can be implemented. Normal operating behaviour is typically defined by expert knowledge and historical data. In industry, statistical process control (Montgomery 2001) has been in use for decades to monitor and control deviations from normal operating behaviour; in pharmaceutical industry it has become known as Process Analytical Technology (Chanda et al. 2015). Also in Chemometrics much research has been devoted to it over the years (Kourti and MacGregor 1995; Westerhuis et al. 2000; Kourti 2005; Challa and Potumarthi 2013).

A large number of tools are available, often based on simple plots of parameter values or functions of parameter values over time. A well-known example is given by the so-called Western Electric Rules (Western Electric Co. 1956), that provide decision rules for detecting out-of-control samples or non-random variation, e.g., a single point falling outside the 3σ limits, two consecutive points on the same side of the mean outside 2σ limits, or a larger number of consecutive points (often seven, or nine) falling at the same side. Many sets of rules exist, all depending on heuristically defined control limits and/or action limits. What type of action needs to be taken depends on the situation; also for a process that is in control one would expect these rules to be activated quite regularly, so the usual approach is to first investigate the matter more closely and only take further action if something is shown to be clearly wrong.

The most well-known example is the so-called Shewhart control chart[1] showing either individual measurement values or averages of groups of samples over time. In what is usually called Phase 1, a subset of the (groups of) samples, known to be "in control", are used to calculate the central value and the so-called upper and lower control limits. These are often taken to be the average and three standard deviations above and below the average, respectively. Such an approach implicitly assumes a normal distribution of the in-control data; obviously this may be an underestimation of the complexity of real-world data. An example where a mixture of normal distributions is used to describe normal operating conditions (NOCs) is given in Thissen et al. (2005). The key is to describe variation that can happen without any need for action. In Phase 2 then, the data of interest are plotted in a chart containing the central value and the control limits, and decision rules like the ones mentioned above can be applied.

Similar procedures can be used to monitor other statistics, such as the range or standard deviation of (groups of) data, but in the majority of cases one focuses on measures of location. Particularly interesting are CUSUM charts (Page 1961), plotting cumulative deviations in either positive or negative direction, and exponentially-weighted moving average (EWMA) charts (Roberts 1959). Both are able to flag cases where individual deviations are relatively small.

The basic ideas will be illustrated here using the *Arabidopsis thaliana* data seen in Sect. 11.1. Since the number of samples is quite big, the measurement series was split up in several batches, a common procedure in many metabolomics experiment. Unfortunately, the experiment was marred by a particularly unlucky series of events (including a broken oil pump and multiple power cuts), leading to quite substantial differences between (and even within) batches. The data are available as the `arabidopsis` data set in the **ChemometricsWithR** package. Here we are focusing on the second metabolite, which is present in all samples in these batches (the first metabolite contains a lot of missing values). We will use the first batch B1 to define NOCs, and then investigate whether we see evidence of changes in batches B2 and B3.

```
> metabNr <- 2
> B1 <- which(arabidopsis.Y$Batch == "B1")
> B23 <- which(arabidopsis.Y$Batch %in% c("B2", "B3"))
> ara.qcc <- qcc(data = arabidopsis[B1, metabNr], type = "xbar.one",
+                newdata = arabidopsis[B23, metabNr],
+                plot = FALSE)
> ara.cusum <- cusum(data = arabidopsis[B1, metabNr],
+                newdata = arabidopsis[B23, metabNr],
+                plot = FALSE)
```

The `qcc` function (from the package with the same name) implements the Shewhart chart for groups of samples (showing group averages in the chart), as well as for individual samples. In the latter case, like here, the `type` argument gets the value `"xbar.one"`. The first argument of the `qcc` and `cusum` functions presents

---

[1] After Walter A. Shewhart, the pioneer in the field working at Bell Labs in the 1920s.

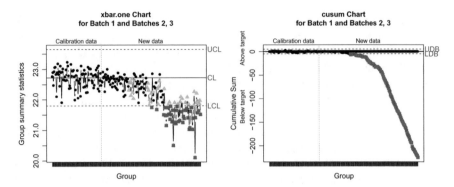

Fig. 11.7 Shewhart (left) and CUSUM (right) charts for the second metabolite in the `arabidopsis` data, batches B1, B2 and B3 only. Batch B1 is used to set the upper and lower control limigs. Warning points are indicated as orange triangles, out-of-control points as red squares. Clearly, stability is not maintained across batches

the data for which the NOCs are known to be satisfied—these will be used to calculate the upper and lower control limits (indicated by UCL and LCL in the figures) if not provided by the user. Argument `newdata` contains the rest of the data that should be included in the plots, typically the most interesting part.

Figure 11.7 is generated by simply plotting the `ara.qcc` and `ara.cusum` objects, including some additional manipulation (not shown) to keep the figures as simple as possible. The Shewhart chart on the left shows that samples deviate more and more from the specifications set up in B1 – initially, the samples are still within the control limits, but then the first warning signs pop up, samples for which too many consecutive values are below the central value. These are indicated with orange triangles. In the last third of the data we see many samples outside the control limits (indicated with red squares), and basically all samples are flagged one way or another. The CUSUM chart on the right shows two statistics at the same time: cumulative positive sums (here almost all zero), as well as cumulative negative sums. The latter show a clear and continuing movement away from the central values. CUSUM charts in general are more sensitive than Shewhart charts—also here deviations are clear more quickly in the CUSUM chart on the right.

Multivariate charts are very similar, but now the statistic that is shown is derived from a multivariate analysis and can be, e.g., a principal component score, or a distance to a nominal value (typically taking the covariance structure into account by using the Mahalanobis distance, Eq. 7.3). The chart showing the distance to a central value is known as the Hotelling T^2 chart, and is probably the most popular of all multivariate control charts. One difference with the Shewhart chart is that since distances are only non-negative, the only relevant control limit is the upper one. In PCA-based charts, one can focus on the subspace covered by the chosen number of PCs, or on the opposite, the residuals (Jackson 1991), completely analogous to the score and orthogonal distances from Fig. 11.6 in the previous section. In all cases, much depends on the scaling, and like with the previous multivariate control charts one should take care to scale all data according to the reference set in the NOCs. In a multivariate context one may be interested in trying to assess which variables contribute to a point being flagged as a warning or outlier point.

The R package **MSQC** (Santos-Fernández 2013) provides several examples. We will show two multivariate control charts for the `arabidopsis` metabolomics data. The first chart is the T^2 control chart. Missing values are not allowed, so we will either have to impute them, or somehow eliminate them. Here we choose the latter option by focusing on metabolites without missing values only. For large numbers of variables the output of the `mult.chart` with argument `type = "t2"` (the "real" T^2 chart) becomes rather lengthy and computing time increases quite dramatically, caused by calculations trying to identify the variables associated with outlyingness. Here we use `type = "chi"` instead, leading to a χ^2-chart with virtually the same results in a fraction of the time.

```
> idx <-
+     which(apply(arabidopsis, 2, function(x) sum(is.na(x)) == 0))
> chi2chart <-
+     mult.chart(arabidopsis[c(B1, B23), idx], type = "chi",
+                Xmv = colMeans(arabidopsis[B1, idx]),
+                S = cov(arabidopsis[B1, idx]))
> abline(v = length(B1) + .5, lty = 2)
>
> MCUSUMchart <-
+     mult.chart(arabidopsis[c(B1, B23), idx], type = "mcusum2",
+                Xmv = colMeans(arabidopsis[B1, idx]),
+                S = cov(arabidopsis[B1, idx]))
> abline(v = length(B1) + .5, lty = 2)
```

This leads to the left panel in Fig. 11.8. Again the difference between the samples from the first batch and the later batches is quite obvious. Multivariate cusum charts (MCUSUM) follow the same principle: they track a univariate statistic that is calculated from multivariate data. Several different flavours exist; **MSQC** implements

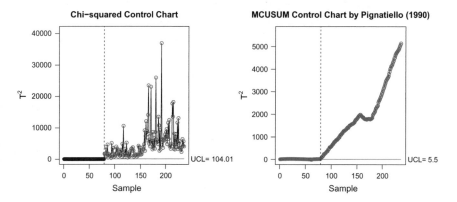

Fig. 11.8 Multivariate control charts for the complete metabolites in the `arabidopsis` data, batches `B1`-`B3` only. NOCs have been defined using batch `B1` (left of the vertical dashed line)

two of them (Crosier 1988; Pignatiello and Runger 1990). For the current data they lead to virtually the same results—the chart is shown in the right panel of Fig. 11.8.

Finally, it is worth mentioning that one can also base multivariate control charts on transformations of the original data, such as the scores of a PCA model, or if additional dependent variables are present, scores of a PLS model. This may eliminate quite a lot of hopefully less relevant variation, leading to fewer false positives while maintaining sensitivity. Also robust forms of PCA can be used, where the effect of single outlying samples will be much smaller and control will focus on general effects.

The control limits are based on heuristics, although they may be calculated using some fancy formulas. Whether one really should take action, perhaps changing process parameters or flagging data points as "unreliable" depends on the situation and the costs of both false positive and false negative alarms. Note also that the data here are quite particular, mainly because of their high dimensionality and large chunks of systematic variations, and may not resemble typical applications for statistical process control. In Sect. 11.7 we will go into methods for batch correction that for these data are probably more appropriate.

11.4   Orthogonal Signal Correction and OPLS

Orthogonal Signal Correction (OSC) was first proposed by Wold and coworkers (Wold et al. 1998) with the aim to remove information from X that is orthogonal to Y. Several different algorithms have been proposed in literature—a concise comparison of several of them has appeared in Svensson et al. (2002). The conclusion of that paper is that OSC in essence does not improve prediction quality per se but rather leads to more parsimonious models, that are easier to interpret. Moreover, the part of X that has been removed before modelling can be inspected as well and may provide information on how to improve measurement quality.

As an example, we will show one form of OSC, called Orthogonal Projection to Latent Structures (OPLS, Trygg and Wold 2002), as summarized in Svensson et al. (2002). Using the weights w and loadings p from an initial PLS model, the variation orthogonal to the dependent variable is obtained and subtracted from the original data matrix:

$$w_\perp = p - \frac{w^T p}{w^T w} w \tag{11.1}$$

$$t_\perp = X w_\perp \tag{11.2}$$

$$p_\perp = \frac{X^T t_\perp}{t_\perp^T t_\perp} \tag{11.3}$$

$$X_c = X - t_\perp^T p_\perp \tag{11.4}$$

The corrected matrix X_c is then used in a regular PLS model. If desired, more orthogonal components can be extracted. This procedure basically comes down to a rotation of PLS components within the selected subspace, followed perhaps by a further dimension reduction, and focuses even more aggressively than standard PLS on the information in the dependent variable. In general, OPLS needs fewer components than PLS.

Let us see how this works out for the gasoline data. From Fig. 8.4 we have concluded that, based on a training set consisting of the odd rows of the `gasoline` data frame, three PLS components are needed. To make things easier, we start by mean-centering the spectra, based on the mean of the training data only. The OSC-corrected matrix is then obtained as follows:

```
> gasolineSC <- gasoline
> gasolineSC$NIR <-
+   scale(gasolineSC$NIR, scale = FALSE,
+         center = colMeans(gasolineSC$NIR[gas.odd, ]))
> gasolineSC.pls <- plsr(octane ~ ., data = gasolineSC, ncomp = 5,
+                        subset = gas.odd, validation = "LOO")
> ww <- gasolineSC.pls$loading.weights[, 1]
> pp <- gasolineSC.pls$loadings[, 1]
> w.ortho <- pp - c(crossprod(ww, pp)/crossprod(ww)) * ww
> t.ortho <- gasolineSC$NIR[gas.odd, ] %*% w.ortho
> p.ortho <- crossprod(gasolineSC$NIR[gas.odd, ], t.ortho) /
+   c(crossprod(t.ortho))
> Xcorr <- gasolineSC$NIR[gas.odd, ] - tcrossprod(t.ortho, p.ortho)
```

Next, a new PLS model is created using the corrected data matrix:

```
> gasolineSC.osc1 <- data.frame(octane = gasolineSC$octane[gas.odd],
+                               NIR = Xcorr)
> gasolineSC.opls1 <- plsr(octane ~ ., data = gasolineSC.osc1,
+                          ncomp = 5, validation = "LOO")
```

Removal of a second OSC component proceeds along the same lines:

```
> pp2 <- gasolineSC.opls1$loadings[, 1]
> w.ortho2 <- pp2 - c(crossprod(ww, pp2)/crossprod(ww)) * ww
> t.ortho2 <- Xcorr %*% w.ortho2
> p.ortho2 <- crossprod(Xcorr, t.ortho2) / c(crossprod(t.ortho2))
> Xcorr2 <- Xcorr - tcrossprod(t.ortho2, p.ortho2)
> gasolineSC.osc2 <- data.frame(octane = gasolineSC$octane[gas.odd],
+                               NIR = Xcorr2)
> gasolineSC.opls2 <- plsr(octane ~ ., data = gasolineSC.osc2,
+                          ncomp = 5, validation = "LOO")
```

Note that the ww vector is the same for every component that is removed (Trygg and Wold 2002). We now can compare the validation curves of the regular PLS model, the PLS model having one orthogonal component removed, and the PLS model with two components removed:

Fig. 11.9 Crossvalidation results for the gasolineSC data (training set only): removal of one or two orthogonal components leads to more parsimonious PLS models

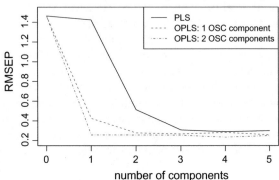

GasolineSC training data (validation)

```
> plot(gasolineSC.pls, "validation", estimate = "CV",
+       ylim = c(0.2, 1.5),
+       main = "GasolineSC training data (validation)")
> lines(0:5, c(RMSEP(gasolineSC.opls1, estimate = "CV"))$val,
+       col = 2, lty = 2)
> lines(0:5, c(RMSEP(gasolineSC.opls2, estimate = "CV"))$val,
+       col = 4, lty = 4)
> legend("topright", lty = c(1, 2, 4), col = c(1, 2, 4),
+        legend = c("PLS", "OPLS: 1 OSC component",
+                   "OPLS: 2 OSC components"))
```

The result is shown in Fig. 11.9. Clearly, the best prediction errors for the two OPLS models are comparable (even slightly better) to the error in the original model using three components, and the optimal values are reached with fewer latent variables.

To do prediction, one has to deflate the new data in the same way as the training data; i.e., one has to subtract the orthogonal components before presenting the data to the final PLS model.

```
> Xtst <- gasolineSC$NIR[gas.even, ]
> t.tst <- Xtst %*% w.ortho
> p.tst <- crossprod(Xtst, t.tst) / c(crossprod(t.tst))
> Xtst.osc1 <- Xtst - tcrossprod(t.tst, p.tst)
> gasolineSC.opls1.pred <- predict(gasolineSC.opls1,
+                          newdata = Xtst.osc1,
+                          ncomp = 2)
```

Predictions for the OPLS model with two OSC components removed are obtained in the same way:

```
> t.tst2 <- Xtst.osc1 %*% w.ortho2
> p.tst2 <- crossprod(Xtst.osc1, t.tst2) / c(crossprod(t.tst2))
> Xtst.osc2 <- Xtst.osc1 - tcrossprod(t.tst2, p.tst2)
> gasolineSC.opls2.pred <- predict(gasolineSC.opls2,
+                          newdata = Xtst.osc2,
+                          ncomp = 1)
```

We can now compare the RMSEP values for the different PLS models:

```
> RMSEP(gasolineSC.pls, newdata = gasolineSC[gas.even, ],
+        ncomp = 3, intercept = FALSE)
[1]   0.2093
> rms(gasolineSC$octane[gas.even], gasolineSC.opls1.pred)
[1] 0.37902
> rms(gasolineSC$octane[gas.even], gasolineSC.opls2.pred)
[1] 0.44888
```

Although the crossvalidation errors do not increase, the prediction of the unseen test data deteriorates quite a bit.

In metabolomics, OPLSDA (Bylesjö et al. 2006), consisting of the combination of PLSDA and OSC, is very popular (Worley and Powers 2013). It has the same advantage as OPLS over PLS: the final model may contain fewer components, although the actual predictions will remain the same (Tapp and Kemsley 2009). OPLSDA applied to a two-class problem, e.g., by definition should contain one component only. However, it would be wrong to see this as an excuse to forego crossvalidation procedures, since it *does* matter how many original PLS components are taken into account in order to arrive at the final OPLSDA model. Moreover, there has always been a tendency in the chemometrics world to attach large meaning to PLS components (scores and loadings)—this may have been true in the original spectroscopic applications based on NIR and UV/Vis data, but in many of the areas in which PLS and friends are being applied now it is very hard to make sense of loading vectors. Very often model inspection is much easier by concentrating on the regression vector: this is after all what new data will be multiplied with when obtaining predictions.

Functions for OPLS and OPLSDA are available in Bioconductor package **ropls** (Thévenot et al. 2015).

11.5   Biomarker Identification

Although there are certainly cases where a multivariate fingerprint is used to predict the class of a new sample, as discussed in the previous paragraph, in many cases the classification per se is not the ultimate goal. We do not need millions of lab equipment to see if someone is healthy or diseased, or a person is a man or a woman. The question very often is: what exactly is different between the classes? The assumption is that the classification model we have built can be interrogated to give us some clues. Important variables are those variables that really influence predictions. If we have used autoscaled data, as is customary with many methods, then the coefficient size is a first clue: coefficients with larger absolute sizes exert more influence. If we know confidence intervals for individual coefficients, e.g., by using the bootstrap methods from Sect. 9.6, then we can also use these and focus on variables that do not include zero in their confidence interval. Note that the variable selection methods discussed in Chap. 10 could be seen in the same light: even though variables are selected purely on the basis of improving estimated prediction errors the final set can be seen as informative—the same holds for penalized regression methods like the lasso that lead to implicit variable selection.

The important distinction that needs to be made here is again the one between cause and effect. Even if we do see a relation, we can never prove that it is causal. Especially with high-dimensional data sets we will always see many correlated variables. If variable selection procedures or model coefficients suggest to focus on a particular subset of the variables, be aware that probably many other subsets exist that are similarly good in terms of prediction but can be interpreted completely differently. In some cases it pays to consider highly correlated variables in groups.

Here we will focus on a metabolomics LC-MS data set containing twenty apple samples (Franceschi et al. 2012), a subset of which is available from the CRAN package **BioMark** (Wehrens and Franceschi 2012a). Ten of the apple samples (each apple leads to one separate sample) were spiked with a mixture of nine chemical compounds, naturally occurring in apples—the goal is to see how successful different methods are in highlighting the differences between the unspiked and spiked apples. The nine compounds have also been measured separately and 5 features that could be unambiguously assigned to spike-in compounds have been identified. The data matrix used here has 197 columns.

Since true differences between the samples are so few, we should not expect unsupervised, global methods like PCA to be very helpful. One approach would be to simply perform t-tests for each variable, and to see how many of the truly different mass peaks are picked up. Alternatively, we could look at the performance of the lasso, or look at the coefficients of a PLSDA model, and it is these three approaches we will concentrate on in the following. The t-test and the lasso have automatic selection rules (based on the chosen level of α and the strength of the L_1 penalty, respectively), but the PLSDA model will lead to a coefficient vector in which all variables will probably have non-zero values. As always, we should first define the optimal values for the number of components in the PLSDA model, and the amount of penalization in the lasso. For simplicity we will fix the number of PLS components to 3 here. First, peak intensities are transformed by taking square roots:

```
> data(spikedApples, package = "BioMark")
> X <- sqrt(spikedApples$dataMatrix)
> Y <- rep(0:1, each = 10)
> apple.df <- data.frame(Y = Y, X = X)
> apple.pls <- plsr(Y ~ X, data = apple.df, ncomp = 5,
+                   validation = "LOO")
> nPLS <- selectNcomp(apple.pls, method = "onesigma")
> apple.lasso <- cv.glmnet(X, Y, family = "binomial")
```

The model coefficients, together with the t statistics for individual features, are first arranged in a matrix:

```
> tvals <- apply(X, 2,
+               function(x) t.test(x[1:10], x[11:20])$statistic)
> allcoefs <-
+   data.frame(studentt = tvals,
+              pls = c(coef(apple.pls, ncomp = nPLS)),
+              lasso = coef(apple.lasso,
+                           lambda = apple.lasso$lambda.1se)[-1])
```

Next, we visualize them in a pairs plot:

```
> N <- ncol(spikedApples$dataMatrix)
> biom <- spikedApples$biom
> nobiom <- (1:N)[-spikedApples$biom]
> pairs(allcoefs,
+       panel = function(x, y, ...) {
+         abline(h = 0, v = 0, col = "lightgray", lty = 2)
+         points(x[nobiom], y[nobiom], col = "lightgray")
+         points(x[biom], y[biom], cex = 2)
+       })
```

The result is shown in Fig. 11.10. The correlation between the t statistics and the PLSDA coefficients is readily apparent: the four largest PLS coefficients correspond with the largest negative t statistics and are all related to true biomarkers. In both cases, the fifth biomarker is not too far behind in the pecking order. The lasso has, as expected, most (in fact: all but five) coefficients equal to zero and only finds one biomarker. For the t statistics and PLS coefficients we would have to define cutoff limits discriminating between potential biomarkers and not-so-interesting variables in order to proceed, but here we will leave it at this point.

The **BioMark** package provides a very simple way to perform analyses like the one above using the get.biom function. Exactly the same results for student's t test and the PLS and lasso models can be obtained in one command:

```
> biomarkerSets <-
+   get.biom(X, factor(Y), type = "coef", ncomp = nPLS,
+            fmethod = c("studentt", "pls", "lasso"))
```

Since the second argument is a two-level factor, the function assumes classification is the goal. The result is a list, with an element for each of the fmethod entries. The "studentt" and "pls" results are simple coefficient vectors, but the "lasso" element is a matrix, with one column for each of the λ penalties considered. If we would have selected multiple PLS components, also the PLS result would have been a matrix.

Apart from variable selection based on coefficient size, **BioMark** implements several other criteria. One is Stability Selection (Meinshausen and Bühlmann 2010; Wehrens et al. 2011), basically repeating a particular selection procedure on different subsets of the data, and retaining only those choices that are more or less consistently present across all results. For methods that do not select a subset themselves, such as PLS, the selection is made on the basis of the coefficient size, so the approach is very much related to the selection process described above. A second one is called

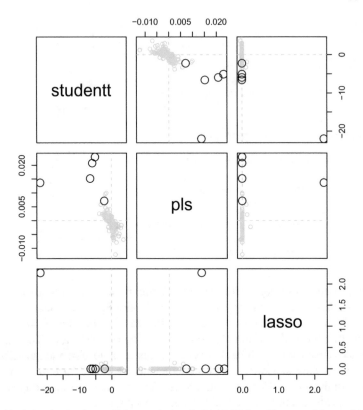

Fig. 11.10 Pairs plot for the coefficients calculated with PLSDA and lasso models, combined with the test statistics from repeated t tests. True biomarkers are indicated in larger black symbols. The t test and PLSDA pick up four out of five biomarkers as the largest coefficients. The lasso selects five non-zero coefficients, the largest of which indeed corresponds to a biomarker

Higher Criticism (Donoho and Jin 2004, 2008; Wehrens and Franceschi 2012b) after an idea originally proposed by John Tukey. This approach basically uses the deviation of a set of p values from the expected uniform behavior under the null distribution to select a cut-off point after which differences are no longer interesting. Neatly summarized by the phrase "the z score of the p value" it assumes that in the large majority of cases the null hypothesis is true, a situation that should apply to many cases in metabolomics, and maybe even in -omics sciences in general. Recently, Generalized Higher Criticism (GHC) was introduced (Barnett et al. 2017) to explicitly take correlations between variables into account.

An example of applying stability selection to PCDA and PLSDA is shown below:

```
> apple.stab <- get.biom(X = X, Y = factor(Y), ncomp = 1:3,
+                        type = "stab", fmethod = c("pls", "pcr"))
> selected.variables <- selection(apple.stab)
> unlist(sapply(selected.variables, function(x) sapply(x, length)))
  pls pcr
1  38  45
2  46  40
3  48  41
```

The two methods (indicated with `"pls"` for PLSDA and `"pcr"` for PCDA, respectively) lead to approximately the same number of variables that consistently show up in the top ten of the model coefficients in terms of size. Looking at the top ten is an arbitrary choice. One can change this using the `biom.options` function; it is also possible to indicate the size of the primary selections by a fraction of the original number of variables.

Just to put the problem into context, it is important to realize that in practice resources for following up on potential biomarkers are often limited. That means that the emphasis usually is on prioritization rather than on selecting the "correct" set. It is quite common to see similar results when comparing lists of biomarkers obtained with different methods—"different" methods are not always very different under the hood. However, it is important to remember that these lists of potential biomarkers are basically hypotheses that should be confirmed or refuted in follow-up experiments. Using domain knowledge to interpret the results in a useful way is essential.

11.6 Calibration Transfer

Imagine a company with high-quality expensive spectrometers in the central lab facilities, and cheap simple equipment on the production sites, perhaps in locations all over the world—you can easily see why a calibration model set up with data from the better instruments (the "master" instruments) may do a better job in capturing the essentials from the data. In general, however, it is not possible to directly use the "good" model for the inferior ("slave") instruments: there will be systematic differences, such as different wavelength ranges and resolutions, as well as some less-clear ones; every measuring device has its own characteristics. Predictions directly using the model from the master instrument, if possible at all, will not always be of a high quality.

Constructing individual calibration models for every instrument separately is of course the best strategy to avoid incompatibilities. However, this may be difficult because calibration samples may not be available, or very expensive. Ideally then, one would want to use the model of the master instrument for the slave spectrometers, too. Several ways of achieving this have been proposed in the literature (de Noord 1994). For one thing, one may try to iron out the differences between the slaves and the master by careful preprocessing, usually including variable selection. In this way, parts of the data that are not too dependent on the instrument—but are still relevant for calibration purposes—can be utilized. This is sometimes called *robust calibration* in chemometrics literature, and is not to be confused with robust regression as is known in statistics.

Another approach has become known under the phrase *calibration transfer* or *calibration standardization* (Wang et al. 1994). The goal is to use a limited number of common calibration samples on the slave instruments to adapt the model developed on the master. This strategy is also useful in cases where instrument response changes over time, or when batch-to-batch differences occur. Although a complete recalibration is always better, it has been reported that an approach using standardization leads to an increase in errors by only 20–60%, which in practice may still be perfectly adequate.

Several variants of calibration transfer have been published (Wang et al. 1995; Bouveresse et al. 1995). The simplest is *direct standardization* (DS, Wang et al. 1994). For a set of samples measured on both instruments, one constructs a transformation F relating the measurements on the master instrument, X_1, to the measurements on the slave X_2:

$$X_1 = X_2 F \tag{11.5}$$

The transformation can be easily obtained by a generalized inverse:

$$F = X_2^+ X_1 \tag{11.6}$$

Alternatives are to use PCR or PLS regression, which are less prone to overfitting. Thus, responses measured on the slave instrument can then be transformed to what they would look like on the primary instrument. Consequently, the calibration model of the primary instrument applies—this approach is also known as backward calibration transfer. The reverse (forward transfer) is also possible: spectra of the high-resolution master instrument can be transformed to be similar to the data measured on the lower-quality slaves. The latter option is usually preferred in on-line situations, where speed is an issue.

To illustrate this, we will use the `shootout` data available from the **ChemometricsWithR** package. It consists of NIR data of 654 pharmaceutical tablets, measured at two Multitab spectrometers. Each tablet should contain 200 mg of (undisclosed) active ingredient. The data have already been divided into training, validation and test sets. The complete wavelength range is from 600 to 1898 nm; we will use the first-derivatives of the area between 1100 and 1700 nm (`calibrate.1`). Let us first build a PLS model for the training data, using the **pls** package:

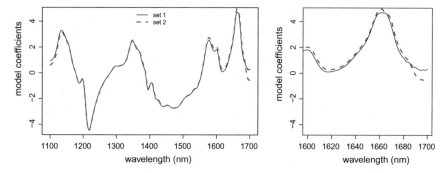

Fig. 11.11 Regression coefficients for the 3-LV PLS models of the NIR shootout data. The right plot zooms in on the area between 1600 and 1700 nm

```
> data(shootout)
> wl <- seq(600, 1898, by = 2)
> indices <- which(wl >= 1100 & wl <= 1700)
> nir.training1 <-
+     data.frame(X = I(shootout$calibrate.1[, indices]),
+                y = shootout$calibrate.Y[, 3])
> mod1 <- plsr(y ~ X, data = nir.training1,
+              ncomp = 5, validation = "LOO")
> RMSEP(mod1, estimate = "CV")
(Intercept)     1 comps     2 comps     3 comps     4 comps
     22.052      17.954       5.827       4.946       4.896
    5 comps
      4.724
```

Three components should be enough. The model based on the spectra measured at the second instrument (`mod2`) is made in the same way, and also requires three latent variables.

Figure 11.11 shows the regression vectors of the two models. Some small differences are visible, in particular in the areas around 1100, 1600 and 1700 nm. As a consequence, `mod1` fares very well in predictions based on spectra measured on instrument 1:

```
> RMSEP(mod1, estimate = "test", ncomp = 3, intercept = FALSE,
+       newdata = data.frame(y = shootout$test.Y[, 3],
+                            X = I(shootout$test.1[, indices])))
[1]   4.974
```

Then again, predictions for data from instrument 2 are quite a bit off:

```
> RMSEP(mod1, estimate = "test", ncomp = 3, intercept = FALSE,
+       newdata = data.frame(y = shootout$test.Y[, 3],
+                            X = I(shootout$test.2[, indices])))
[1]   9.983
```

The average error for predictions based on data from instrument 2 is twice as large— maybe surprising, because the model coefficients do not seem to be very different.

Keep in mind, though, that the intercept is usually not shown in these plots—in this case, a difference of more than twenty units is found between the intercepts of the two models.

Now suppose that five samples (numbers 10, 20, ... 50) are available for standardization purposes: these have been measured at both instruments. Let us transform the data measured on instrument 2, so that the model of instrument 1 can be applied. Now we can use an estimate of transformation matrix F to make the data from the two instruments comparable:

```
> recalib.indices <- 1:5 * 10
> F1 <- ginv(shootout$calibrate.2[recalib.indices, indices]) %*%
+    shootout$calibrate.1[recalib.indices, indices]
> RMSEP(mod1, estimate = "test", ncomp = 3, intercept = FALSE,
+        newdata = data.frame(y = shootout$test.Y[, 3],
+          X = I(shootout$test.2[, indices] %*% F1)))
[1]  4.485
```

Immediately, we are in the region where we expect prediction errors of mod1 to be. Matrix F1 can be visualized using the contour function, which sheds some light on the corrections that are made:

```
> levelplot(F1, contour = TRUE)
```

The result is shown in Fig. 11.12. All numbers are quite close to zero. Horizontal bands correspond with columns in F1, containing the corrections for the spectra of the second instrument X_2. The largest changes can be found in the final thirty or so variables corresponding to the area between 1640 and 1700 nm, not surprisingly the area where the largest differences in regression coefficients are found as well.

Because spectral variations are often limited to a small range, it does not necessarily make sense to use a complete spectrum of one instrument to estimate the response at a particular wavelength at the other instrument. Wavelengths in the area of interest are much more likely to have predictive power. This realization has led to *piecewise direct standardization* (PDS, Wang et al. 1994), where only measurements $x_{i-k}, \ldots, x_i, \ldots, x_{i+k}$ of one spectrometer are used to predict x_i at the other spectrometer. This usually is done with PLS or PCR. The row-wise concatenation of these regression vectors, appropriately filled with zeros to obtain the correct overall dimensions, will then lead to F, in the general case a banded diagonal matrix.

An obvious disadvantage of the method is that it takes many separate multivariate regression steps. Moreover, one needs to determine the window size, and information is lost at the first and last k spectral points. Nevertheless, some good results have been obtained with this strategy (Wang et al. 1994).

Methods that directly transform the model coefficients have been described as well, but results were disappointing (Wang et al. 1992). Apart from this, a distinct disadvantage of transforming the model coefficients is that the response variable y is needed. The DS and PDS approaches described above, on the other hand, can be applied even when no response information is available: in Eq. 11.6 only the spectral data are used.

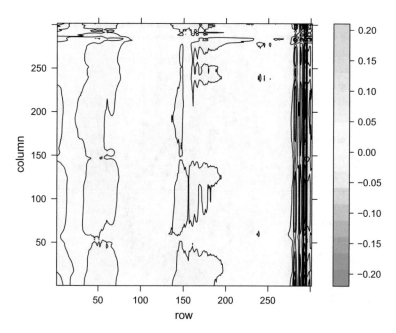

Fig. 11.12 Contour lines of the transformation matrix `F1`, mapping spectral data of one NIR instrument to another

11.7 Batch Correction

In the previous section we tried to make two analytical instruments comparable, in the sense that data measured on one instrument could be analyzed with a model created using data from the other instrument. This works for relatively stable technologies like NIR spectroscopy, but if one is really looking for maximal sensitivity and resolution, even data measured on the same instrument may be not completely comparable. A case in point is mass spectrometry. Large numbers of samples are typically measured in batches, in between which maintenance or cleaning may take place. As a result, batches often show systematic differences, which may or may not be large enough to influence the result. In the area of (MS-based) metabolomics, for example, batch correction is a more or less standard data preprocessing step. The first question obviously is: do we see batch effects in our data? As so often, PCA is very helpful here, the reason being that batch effects, even when small, are typically occurring very consistently across all samples and therefore are likely to show up in global visualization methods like PCA.

Again we look at the `arabidopsis` data set introduced in Sect. 11.1. In this case, missing values have been imputed by column means. From the PCA scoreplot in Fig. 11.13 it is clear that batches 1 and 2 stand out from the rest; for some other batches we see reasonably good overlap. Batch 8 is an interesting case: one of the power cuts took place in the middle of measuring this batch, and one can clearly

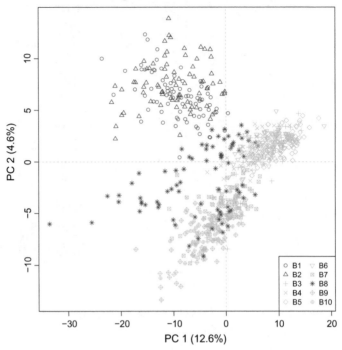

Fig. 11.13 PCA visualization of autoscaled metabolomics data from a ten-batch LC-MS measurement of *Arabidopsis thaliana* samples. Batch effects are clearly visible—in particular, batches 1, 2 and 8 (highlighted) stand out. The average Bhattacharyya distance between the batches is shown in the plot title

see two point clouds: it could be a good idea to split this batch into two sub-batches before applying batch correction methods (we will not do so here). Batch-to-batch differences can be quantified using the Bhattacharyya distance which measures the distance between two normal distributions (note that the batch PCA scores do resemble normal distributions). It is defined by

$$D_B = \frac{1}{8}(\mu_1 - \mu_2)^T \Sigma^{-1}(\mu_1 - \mu_2) + \frac{1}{2}\left(\frac{\det \Sigma}{\sqrt{\det \Sigma_1 \det \Sigma_2}}\right)$$

where μ_1, μ_2, Σ_1 and Σ_2 are the means and covariance matrices of the two distributions, and

$$\Sigma = \frac{\Sigma_1 + \Sigma_2}{2}.$$

Rather than using the full data, we use the PCA scores in, e.g., two dimensions, which leads to pleasantly simple covariance matrices and easy calculations. If no batch effects are present, the batches should overlap and the distances should be very close to zero. Figure 11.13 contains the average between-batch distance in the figure title.

Many different techniques can be applied for batch correction (Dunn et al. 2011; Hendriks et al. 2011). Typically, one concentrates on quality control (QC) samples, injected several times in each batch. These can then be used to assess average response values within each batch, and even to model and later correct trends within a batch. The top panel in Fig. 11.14 presents data for the second metabolite in the `arabidopsis` data set also discussed in Sect. 11.3:

```
> ara.df <- cbind(data.frame(X = arabidopsis[, 2]),
+                 arabidopsis.Y)
> ref.idx <- ara.df$SCode == "ref"
> plot(X ~ SeqNr, data = ara.df, ylim = c(20, 24),
+      col = as.numeric(ref.idx) + 1,
+      pch = c(1, 19)[as.numeric(ref.idx) + 1],
+      xlab = "Injection number", ylab = "Intensity (log-scaled)",
+      main = paste("Metabolite 2 before correction"))
> batch.lims <- aggregate(ara.df$SeqNr,
+                         by = list(ara.df$Batch),
+                         FUN = range)$x
> abline(v = batch.lims[-1, 1] - 0.5, lty = 2)
```

There are clear differences between the batches, and in several cases also clear trends within batches. After correcting using the straight lines through the QC samples, the behaviour is much more uniform, as indicated in the bottom panel. For this example, the correction can be easily performed in a few line of code. First we estimate the regression lines through the QC samples for the individual batches. These correspond to the values that would be expected for QC samples at these locations, indicated in Fig. 11.14 as red lines:

```
> BLines <- lm(X ~ SeqNr * Batch, data = ara.df,
+              subset = SCode == "ref")
> ara.df$QCpredictions <- predict(BLines, newdata = ara.df)
> for (ii in levels(ara.df$Batch))
+    lines(ara.df$SeqNr[ara.df$Batch == ii],
+          predict(BLines, newdata = ara.df[ara.df$Batch == ii, ]),
+          col = 2)
```

Next, we simply subtract these QC predictions from the actual data, and add the global mean to get the data to a "normal" level:

```
> ara.df$corrected <-
+    ara.df$X - ara.df$QCpredictions + mean(ara.df$X)
```

Plotting the corrected values then leads to the bottom panel in Fig. 11.14. By definition all lines through the QCs are horizontal and at the same level, but also the variation in the study sample intensities has decreased markedly. A particular advantage of this method is that it is capable of handling missing values: only if too few QC samples are available to estimate the within-batch trends or batch averages the correction

Metabolite 2 before correction

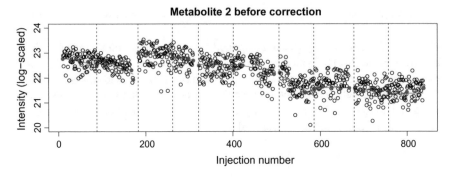

Metabolite 2 after correction

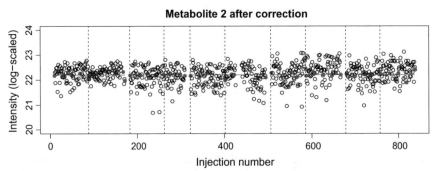

Fig. 11.14 Intensities of one of the metabolites in the `arabidopsis` data set. Batches are indicated with vertical dashed lines; QC samples—on which the straight lines are based—are indicated with filled circles. To facilitate easy comparison, the vertical scales of the two plots are equal

cannot be performed. Note that the PCA score plots and the associated between-batch distance measures cannot handle missing values, and these have been imputed by column means in order to least affect the plots and the numerical results.

Limiting ourselves to those metabolites showing fewer than 50 missing values, and estimating linear trends within batches using not only the QC samples but also the study samples, we can easily obtain a full corrected data matrix:

```
> correctfun <- function(x, seqnr, batch) {
+    huhn.df <- data.frame(x = x, seqnr = seqnr, batch = batch)
+    blines <- lm(x ~ seqnr * batch, data = huhn.df)
+    huhn.df$qcpred <- predict(blines, newdata = huhn.df)
+    huhn.df$x - huhn.df$qcpred
+ }
>
> nna.threshold <- 50
> x.idx <- apply(arabidopsis, 2, function(x) sum(is.na(x)) < 50)
> correctedX <- apply(arabidopsis[, x.idx], 2, correctfun,
+                     seqnr = arabidopsis.Y$SeqNr,
+                     batch = arabidopsis.Y$Batch)
```

Fig. 11.15 The
`arabidopsis` data from
Fig. 11.13 after batch
correction. No particular
batch structure can be
observed, in agreement with
the very low Bhattacharyya
distance shown in the plot
title

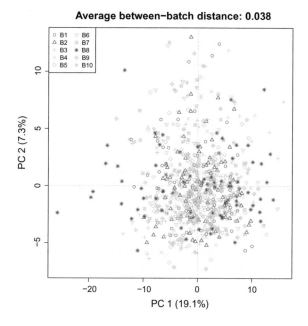

Well, "full": only 163 metabolites satisfy the missing-values criterion. The PCA
scoreplot of this corrected matrix is shown in Fig. 11.15. In comparison to Fig. 11.13
the batches are much more overlapping, also indicated by the much lower between-
batch distance in the plot title.[2]

Not only linear but also non-linear approximations for trends, often based on
smoothing techniques like the loess have been used—note that when the number
of QC samples is low, non-linear smoothing like the `loess` function mentioned in
Chap. 3 is not without risks. Alternatively, one may use all samples, including QC as
well as study samples to estimate averages and trends within batches (Wehrens et al.
2016), provided that the injection order is properly randomized. This leads to more
variation, so if one applies smoothing techniques one should be careful not to follow
local variations too quickly, but has the advantage that many more points become
available. Since missing values are a common problem in metabolomics, including
more samples leads to more chances for succesful batch correction.

So far, we have corrected each variable individually, and inspection of the correction factors shows that this indeed is necessary: apparently batch effects differ
for individual metabolites. Still, we could hypothesize that a multivariate correction
would be able to benefit from correlations between metabolites. A method that does

---

[2]One may object here: many metabolites were not taken into account because they contained too
many missing values. However, the between-batch distance for the subset of the original data matrix
corresponding to the columns that could be corrected is even larger than the one of the full matrix:
3.868. This may be surprising at first, but on second thought it makes sense: batch effects are most
visible in columns that contain few missing values.

this is called "Removal of Unwanted Variation" (RUV), and implemented in the R package **RUVSeq** (Risso et al. 2014). Originally devised for genomics data, it is based on a PCA decomposition of the data from the QC samples (Gagnon-Bartsch and Speed 2012; Livera et al. 2015). Projecting the study samples in this subspace gives an idea of the unwanted variation present in the data, and an opportunity to remove it. The only choice that the user needs to make is the dimensionality of the subspace – it has been reported that values in the range of three to ten PCs lead to very similar results (Wehrens et al. 2016). The disadvantage of this approach is that missing values need to be imputed before the method can be applied. For the visualisation of batch effects in PCA score plots we used column means to impute missing values (the correction itself would not use these imputed values), but to do so here would be quite wrong. One would rather use a value like the smallest intensity measured in the data set (Wehrens et al. 2016) or even draw from estimated log-normal districtions (Uh et al. 2008). Note that imputation by zeroes would be among the worst possible choices since actual values typically are very far away from zero.

As an example, let's apply the RUV method to the column subset with fewer than 50 missing values, imputing them by column minima.

```
> X <- arabidopsis[, x.idx]
> na.idx <- is.na(X)
> X[na.idx] <- min(X, na.rm = TRUE)
```

The RUVs function uses the scIdx argument to specify which samples are the actual replicates. Here, these are the QC samples. For the correction itself we use a three-dimensional PCA subspace, indicated by the k = 3 argument:

```
> idx <- which(arabidopsis.Y$SCode == "ref")
> replicates.ind <-
+   matrix(-1, nrow(X) - length(idx) + 1, length(idx))
> replicates.ind[1, ] <- idx
> replicates.ind[-1, 1] <- (1:nrow(X))[-idx]
> X.RUVcorrected <-
+   t(RUVs(x = t(X), cIdx = 1:ncol(X), k = 3,
+          scIdx = replicates.ind, isLog = TRUE)$normalizedCounts)
> X.RUVcorrected[na.idx] <- NA
```

Note that the data matrix is transposed in the call to RUVs – the result of the fact that the methodology was originally developed for gene expression data, where columns of a data matrix usually refer to samples rather than to variables. The result of the calculation should be transposed, too, to get back to our usual conventions. The scIdx argument is used to indicate which samples are the QC samples. For more information on its exact definition consult the manual page of RUVs. The last line reinserts the missing values. This corrected data matrix leads to an average Bhattacharyya distance of 0.032, slightly smaller than our series of univariate corrections.

11.8 Multivariate Curve Resolution

In Multivariate Curve Resolution (MCR, sometimes indicated with Alternating Least Squares, ALS, or even MCR-ALS), one aims at decomposing a data matrix in such a way that the individual components are corresponding directly to chemically relevant characteristics, such as spectra and concentration profiles. In contrast to PCA, no orthogonality is imposed. This comes at a high price, however: in PCA, the orthogonality constraint ensures that only one linear combination of original variables gives the optimal approximation of the data (up to the sign ambiguity). In MCR, usually a band of feasible solutions is obtained. On the other hand, the orthogonality constraint in PCA prevents one from direct interpretation of the PCs: real, "pure" spectra will almost never be orthogonal. This direct interpretation is the goal of MCR.

Monitoring a reaction by some form of spectroscopy is a classical example: during the reaction products are formed which may, in turn, react to form other compounds. The concentrations of the starting compounds go down over time, those of the end products go up, and those of the intermediates first go up and finally go down again. Some of the components may be known to be present, others may be unexpected and their spectra unknown. Another example, often encountered, is given by data from HPLC-UV analyses: the sample is separated over a column and at certain points in time the UV-Vis spectra are measured. Again, this results in concentration profiles over time for all compounds in the sample, as well as the associated pure spectra.

An example of such a data set is `bdata`. It consists of UV measurements at 73 wavelengths, measured at 40 time points. Two data matrices are available; at this moment we will concentrate on the first. The sample is a mixture of three compounds, two of which are diazinon and parathion-ethyl, both organophosphorus pesticides (Tauler et al. 1996). A perspective plot of the data is shown in Fig. 11.16.

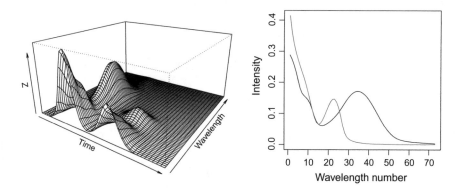

Fig. 11.16 The HPLC-UV data set, a mixture of two compounds eluting at different time points. The left panel shows a perspective plot of the first data matrix; the right panel shows the spectra of the two compounds

11.8.1 Theory

The basis of the method is laid in a seminal paper by Lawton and Sylvestre (Lawton and Sylvestre 1971), but the real popularity within the chemometrics community came two decades later (Tauler 1995; de Juan et al. 2000). In the ideal case, only the number of components and suitable initial estimates for either concentration profiles or pure spectra should have to be provided to the algorithm. The data matrix typically contains the spectra of the mixture at several time points, and can be decomposed as

$$X = CS^T + E \tag{11.7}$$

where C is the matrix containing the "pure" concentration profiles, S contains the "pure" spectra and E is an error matrix. The word "pure" is quoted since there is rotational ambiguity: CR^{-1} and RS^T will, for any rotation matrix R, lead to the same approximation of the data:

$$X = CS^T = CR^{-1}RS^T \tag{11.8}$$

Whether it is possible to identify one particular set of CR^{-1} and RS^T as better than others depends on the presence of additional information. This often takes the form of constraints. If C indeed corresponds to a concentration matrix, it can only contain non-negative elements. Such a non-negativity constraint is applicable for many forms of spectroscopy as well, like a number of other constraints that will be briefly treated below.

   The decomposition of Eq. 11.7 is usually performed by repeated application of multiple least squares regression—hence the name ALS. Given a starting estimate of, e.g., C, one can calculate the matrix S that minimizes the residuals E in Eq. 11.7, which in turn can be used to improve the estimate of C, etcetera.

$$\hat{S} = X^T C (C^T C)^{-1} = X^T C^+ \tag{11.9}$$

$$\hat{C} = X S (S^T S)^{-1} = X \left(S^T\right)^+ \tag{11.10}$$

Equations 11.9 and 11.10 alternate until no more improvement is found or until the desired number of iterations is reached.

11.8.2 Finding Suitable Initial Estimates

The better the initial estimates, the better the quality of the final results—moreover, the MCR algorithm will usually need fewer iterations to converge. Therefore it pays to invest in a good initialization. Several strategies have been proposed. Perhaps the most simple, conceptually, is to calculate the ranks of submatrices, a procedure that has become known in chemometrics as Evolving Factor Analysis (EFA, Maeder 1987). The original proposal was to start with a small submatrix, and monitor the size of the eigenvalues upon adding more and more columns (or rows) to the data matrix; later approaches apply a moving window, e.g., Evolving Windowed Factor Analysis (EWFA, Keller and Massart 1991). Several other methods such as the Orthogonal Projection Approach (OPA, Sanchez et al. 1994) and SIMPLISMA (Windig and Guilment 1991) are not based on SVD, but on dissimilarities between the spectra. These typically return the indices of the "purest" variables or objects. Here, we focus on EFA and OPA.

11.8.2.1 Evolving Factor Analysis

EFA basically keeps track of the number of independent components encountered in the data matrix upon sequentially adding columns (often corresponding to time points). This is especially useful in situations where there is a development over time, such as occurs when monitoring a chemical reaction—the reaction proceeds in a number of steps and it can be interesting to see when certain intermediates are formed, and what the order of the formation is. In this context, the result of an MCR consists of the pure spectra of all species, and their concentration profiles over time.

The most basic implementation of EFA takes a data matrix and the desired number of components as arguments. A basic R implementation could be the following:

```
> efa <- function(x, ncomp) {
+   nx <- nrow(x)
+   Tos <- Fros <- matrix(0, nx, ncomp)
+   for (i in 3:nx)
+     Tos[i, ] <- svd(scale(x[1:i, ], scale = FALSE))$d[1:ncomp]
+   for (i in (nx-2):1)
+     Fros[i, ] <- svd(scale(x[i:nx, ], scale = FALSE))$d[1:ncomp]
+
+   Combos <- array(c(Tos, Fros[, ncomp:1]), c(nx, ncomp, 2))
+   list(forward = Tos, backward = Fros,
+        pure.comp = apply(Combos, c(1, 2), min))
+ }
```

In the forward pass, starting from a submatrix containing only three rows, the singular values of iteratively growing matrices are stored. The backward pass does the same, but now the growth starts at the end and is in the backward direction. The "pure" profiles then are obtained by combining the forward and backward traces, where it is assumed that the first compound to come up is also the first one to disappear. Should one be interested in initial estimates of the other dimension, it suffices to present a transpose matrix as the first argument.

For the HPLC-UV data mentioned above, the forward and backward passes give the result shown in Fig. 11.17:

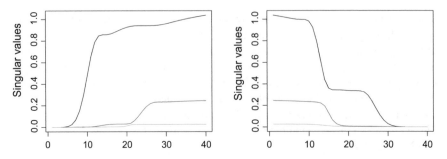

Fig. 11.17 Forward (left plot) and backward (right) traces of EFA on the HPLC-UV data

```
> X <- bdata$d1
> X.efa <- efa(X, 3)
> matplot(X.efa$forward, type = "l", ylab = "Singular values", lty=1)
> matplot(X.efa$backward, type = "l", ylab = "Singular values", lty=1)
```

In the left plot, showing the forward pass of the EFA algorithm, the first compounds starts to come up at approximately the fifth time point. The second and third come in at about the thirteenth and nineteenth time points. In the backward trace we see similar behaviour. Estimates of the "pure" traces are obtained by simply taking the minimal values—we combine the first component in the forward trace with the last component in the backward trace, etcetera.

```
> matplot(X.efa$pure.comp, type = "l", lty = 1,
+         xlab = "wavelength number", ylab = "response")
> legend("topright", legend = paste("Comp", 1:3),
+        lty = 1, col = 1:3, bty="n")
```

The result is shown in Fig. 11.18. Although the elution profiles do not yet resemble neat chromatographic peaks, they are good enough to serve as initial guesses and start the MCR iterations. In some cases, logarithms are plotted rather than the singular values themselves—this can make it easier to see when a compound starts to go up.

11.8.2.2 OPA—The Orthogonal Projection Approach

Rather than estimating concentration profiles, where the implicit assumption is that a compound that comes up first is also the first to disappear, one can also try to find wavelengths that only lead to absorption for one particular compound in the mixture. Methods like the Orthogonal Projection Approach (Sanchez et al. 1994) and SIMPLISMA (Windig and Guilment 1991) have been developed exactly for that situation. Put differently, they focus on finding time points in the chromatograms in which the spectra are most dissimilar. Several related methods are published as well—an overview is available in Sanchez et al. (1996). Here, we will treat OPA only.

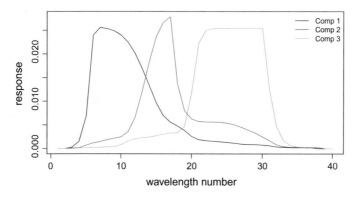

Fig. 11.18 Estimated pure traces using EFA for the HPLC-UV data

The key idea is to calculate dissimilarities of all spectra with a set of reference spectra. Initially, the reference set contains only one spectrum, usually taken to be the average spectrum. In every iteration except the first, the reference set is extended with the spectrum that is most dissimilar—only in the first iteration, the reference is replaced rather than extended by the most dissimilar spectrum. As a dissimilarity measure, the determinant of the crossproduct matrix of Y_i is used:

$$d_i = \det(Y_i^T Y_i) \tag{11.11}$$

where Y_i is the reference set, augmented with spectrum i:

$$Y_i = [Y_{\text{ref}}\ y_i] \tag{11.12}$$

Several scaling issues should be addressed; usually the spectra in the reference set are scaled to unit length (Sanchez et al. 1996), and that is the convention we will also use.

A version of opa is available in the **alsace** package. Just like efa, opa takes a data matrix and the desired number of components as arguments—in addition, the user can supply one or several components that will be used as the reference spectra, the starting point of the OPA algorithm. For the HPLC-UV data, this leads to

```
> X.opa <- opa(X, 3)
```

In this case, the pure spectra rather than the pure elution profiles are obtained as the columns in X.opa. The result is shown in Fig. 11.19. Clearly, the first component represents something very strange and is probably not a valid spectrum. The other two components look better.

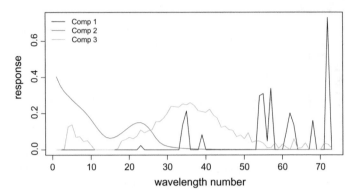

Fig. 11.19 Pure spectra obtained using OPA for the HPLC-UV data set

11.8.3 Applying MCR

The initial estimates of either elution profiles or pure spectra can be used to initiate the MCR iterations of Eqs. 11.9 and 11.10. The simplest possible form of MCR is based on a combination of only a couple of functions. The first, `normS`, takes a set of spectra and normalizes them to unit length:

```
> ## spectra are stored in the rows of the matrix here
> normS <- function(S) {
+    sweep(S,
+          1,
+          apply(S, 1, function(x) sqrt(sum(x^2))),
+          FUN = "/")
+ }
```

Functions `getS` and `getC` take the data to estimate spectra given profiles, or the other way around:

```
> getS <- function(data, C) {
+    normS(ginv(C) %*% data)
+ }
> getC <- function(data, S) {
+    data %*% ginv(S)
+ }
```

The `mcr` function determines which of these to call first, and then continues to call both alternatingly, using updated versions of the estimated spectra or profiles. This continues until the maximum number of iterations is reached, or until convergence:

```
> mcr <- function(x, init, what = c("row", "col"),
+                  convergence = 1e-8, maxit = 50) {
+   what <- match.arg(what)
+   if (what == "col") {
+     CX <- init
+     SX <- getS(x, CX)
+   } else {
+     SX <- normS(init)
+     CX <- getC(x, SX)
+   }
+
+   rms <- rep(NA, maxit + 1)
+   rms[1] <- sqrt(mean((x - CX %*% SX)^2))
+
+   for (i in 1:maxit) {
+     CX <- getC(x, SX)
+     SX <- getS(x, CX)
+
+     resids <- x - CX %*% SX
+     rms[i+1] <- sqrt(mean(resids^2))
+     if ((rms[i] - rms[i+1]) < convergence) break;
+   }
+
+   list(C = CX, S = SX, resids = resids, rms = rms[!is.na(rms)])
+ }
```

Depending on the nature of the initialization, the algorithm starts by estimating pure spectra (input parameter what == "col") or elution profiles (what == "row"). For the Moore-Penrose inverse we again use function ginv from the **MASS** package. The RMS error for the initial estimate is calculated and the iterations are started. The algorithm stops when the improvement is too small. Alternatively, the algorithm stops when a predetermined number of iterations has been reached.

The result of applying this algorithm is visualized in Fig. 11.20:

```
> X.mcr.efa <- mcr(X, X.efa$pure.comp, what = "col")
> matplot(X.mcr.efa$C, type = "n",
+         main = "Concentration profiles", ylab = "Concentration")
> abline(h = 0, col = "lightgray")
> matlines(X.efa$pure.comp * 5, type = "l", lty = 2, col = 1:3)
> matlines(X.mcr.efa$C, type = "l", col = 1:3, lty = 1, lwd = 2)
> legend("topright", legend = paste("Comp", 1:3),
+         col = 1:3, lty = 1, bty = "n")
> matplot(t(X.mcr.efa$S), col = 1:3, type = "l", lty = 1,
+         main = "Pure spectra", ylab = "Intensity")
> abline(h = 0, col = "lightgray")
> legend("topright", legend = paste("Comp", 1:3), lty = 1,
+         bty = "n", col = 1:3)
```

The original concentration profile estimates from EFA have been indicated in dahsed lines in the left plot. Clearly, the new estimates are much better. The peak shapes now are close to what one should expect from a chromatographic separations. There are also a couple of problems: all three compounds seem to have at least two peaks, and some of the concentration values are negative. The corresponding estimates of

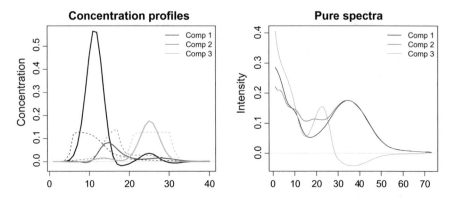

Fig. 11.20 Estimates of concentration profiles (left) and spectra of pure compounds (right) for the HPLC-UV data with MCR. The dashed profiles in the left plot, blown up by a factor of five to show them more clearly, indicate the initialization values from EFA

the pure spectra are shown on the right. Compared to the "true" spectra shown in Fig. 11.16, we see that components 1 and 3 do rather well (although also here we see negative values). Component 2, corresponding to the component with the lowest concentration estimates, seems to be a false positive, or corresponds to an unknown contamination.

The quality of the model can be assessed by looking at the succession of RMS values during the MCR iterations:

```
> X.mcr.efa$rms
[1] 0.02185517 0.00013678 0.00013658 0.00013658
> X.mcr.opa$rms
[1] 0.00487332 0.00345439 0.00014360 0.00013658 0.00013658
```

Both the EFA- and OPA-based models end up with the same error value; note that the initial guess provided by OPA is already very close to the final result. Although both models achieve the same RMS error, they are not identical: this is a result of the rotational ambiguity, where there is a band of solutions with similar or identical quality.

11.8.4 Constraints

One remedy for the rotational ambiguity is to use constraints. From the set of equivalent solutions, only those are considered relevant for which certain conditions hold. In this particular case two constraints are immediately obvious: a non-negativity constraint can be applied to both concentration profiles and spectra, and in addition the concentration profiles can be thought to be unimodal—compounds show a unimodal distribution across the chromatographic column. A non-negativity constraint can be crudely implemented by inserting lines like

```
> SX[SX < 0] <- 0
```

and

```
> CX[CX < 0] <- 0
```

in the `mcr` function, but a better way to do this is to use non-negative least squares. This (as well as several other constraints) has been implemented in the `als` function of package **ALS**. Again, the function requires initial estimates of either C or S. Several ways of normalizing spectra are available—here, we choose `normS = 0.5` which corresponds to scaling the spectra to vectors of unit length. Let us see what this leads to:

```
> X.als.efa <-  als(CList = list(X.efa$pure.comp),
+                   PsiList = list(X), S = matrix(0, 73, 3),
+                   nonnegS = TRUE, nonnegC = TRUE,
+                   normS = 0.5, uniC = TRUE)
Initial RSS 1.6477
Iteration (opt. S): 1, RSS: 1.3928, RD: 0.15471
Iteration (opt. C): 2, RSS: 0.00071595, RD: 0.99949
Iteration (opt. S): 3, RSS: 0.00039749, RD: 0.44481
Iteration (opt. C): 4, RSS: 0.00022507, RD: 0.43378
Iteration (opt. S): 5, RSS: 0.00020012, RD: 0.11084
Iteration (opt. C): 6, RSS: 0.0001692, RD: 0.15453
Iteration (opt. S): 7, RSS: 0.00016206, RD: 0.042201
Iteration (opt. C): 8, RSS: 0.00015392, RD: 0.050228
Iteration (opt. S): 9, RSS: 0.00015045, RD: 0.022493
Iteration (opt. C): 10, RSS: 0.00014986, RD: 0.0039409
Iteration (opt. S): 11, RSS: 0.00014783, RD: 0.013575
Iteration (opt. C): 12, RSS: 0.00014789, RD: -0.00046161
Initial RSS / Final RSS = 1.6477 / 0.00014789 = 11141
```

The output shows the initial RSS, which in this case—since we specify zeros as the initial estimate of S—equals the sum of squares in X. After estimating S, the RSS value has decreased to 0.533. The "RD" in the output signifies the improvement in the corresponding step: using the first estimate for the pure spectra to estimate concentration profiles virtually eliminates the fitting error. After ten iterations, the algorithm stops because there is no further improvement.

The non-negativity constraints for both spectra and diffusion profiles are given by the `nonnegS` and `nonnegC` arguments; `optS1st = TRUE` indicates that the first equation to be solved is Eq. 11.9—giving S0 as an argument is therefore not necessary, although it is necessary to provide a (dummy) matrix of the correct size. The unimodality constraint is indicated with `uniC = TRUE`. The results are shown in Fig. 11.21. The two initializations lead to very similar models in terms of elution profiles, although the spectra are quite different. Note that the order of the components is not the same. Here, the RSS value of the OPA-based model is slightly better: 9.4×10^{-5} vs. 1.48×10^{-4}.

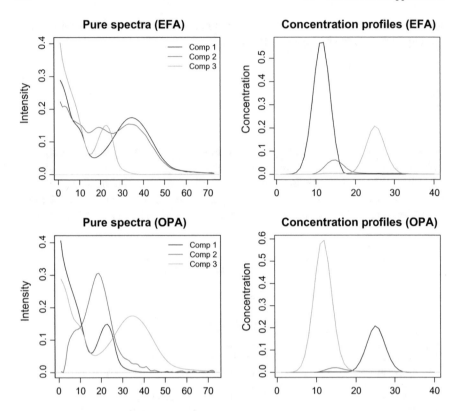

Fig. 11.21 Results for the HPLC-UV data with non-negativity and unimodality constraints: the top row shows the fit after initialization with EFA, the bottom line with OPA

As already stated, constraints are a way to bring in prior knowledge and to limit the number of possible solutions to the chemically relevant ones. Apart from the non-negativity and unimodality constraints encountered in the previous section, several others can be applied. An important example is *selectivity*: in some cases one knows that certain regions in a particular spectrum do not contain peaks from one compound. This forces the algorithm to assign any signal in that region to spectra of other components. Knowledge of mass balances in chemical reactions can lead to *closure* constraints, indicating that the sum of certain concentrations is constant.

The most stringent constraint is to impose an explicit model for one or even both of the data dimensions. One example can be found in the area of diffusion-ordered spectroscopy (DOSY), a form of NMR in which proton patterns of compounds in a mixture are separated on the basis of diffusion coefficients. Theoretical considerations lead to the assumption that the diffusion profiles follow an exponential curve (Stilbs 1981, 1987). Indeed, so-called single-channel algorithms for interpreting DOSY data concentrate on fitting mono- or bi-exponentials to individual variables (Huo et al. 2003). Such data may conveniently be tackled with MCR, where the diffusion profiles are fit using exponential curves. This is reported to lead to more robust results than less stringent constraints (Huo et al. 2004). An extension of the **ALS** package, **TIMP**, allows to do this in R as well. A well-documented example of the use of **TIMP** in the realm of GC-MS data, where individual peaks are represented by generalized normal distributions, can be found in Mullen et al. (2009).

11.8.5 Combining Data Sets

In the classical application, MCR-ALS leads to estimates of pure spectra and pure concentration profiles, given a matrix of several measurements of the mixture. Extensions are possible in cases where either one mixture is studied with different measurement methods, or where several mixtures containing the same components are studied (Munoz and de Juan 2007). In the first case, the concentration profiles of the individual components C are the same, but it is possible to estimate the pure spectra for two or more spectroscopic techniques:

$$[X_1|X_2|...|X_n] = C\left[S_1^T|S_2^T|...|S_n^T\right] \tag{11.13}$$

In the other situation one assumes that the constituents of the different samples are the same, but the concentrations are not:

$$\begin{bmatrix} X_1 \\ X_2 \\ ... \\ X_n \end{bmatrix} = \begin{bmatrix} C_1 \\ C_2 \\ ... \\ C_n \end{bmatrix} S^T \tag{11.14}$$

An example of an application where common spectra and distinct concentration profiles are estimated is available from the `als` manual page. This particular form of MCR allows one to quantify compounds in the presence of unknown interferents: one of the additional data matrices is then the outcome of a measurement of a known quantity of the compound of interest. Because of the linearity of the response in most forms of spectroscopy, it is then possible to relate the concentration in the mixture to that of the standard.

The HPLC-UV data provide a way to see how this works: a second data matrix is present, containing the same three compounds. Since the variability of the chromatographic separation is much larger than the variability in spectral response, we assume common spectra, and will estimate two sets of concentration profiles. This can be done with the following code:

```
> C0 <- matrix(0, 40, 3)
> X2.als.opa <-
+   als(CList = list(C0, C0),
+       PsiList = list(bdata$d1, bdata$d2),
+       S = X.opa, normS = 0.5,
+       nonnegS = TRUE, nonnegC = TRUE,
+       optS1st = FALSE, uniC = TRUE)
```

We can assess how well the true spectra are approximated by the MCR results by looking at the correlations:

```
> cor(X.als.opa$S, cbind(c(bdata$sp1), c(bdata$sp2)))
        [,1]    [,2]
[1,]  0.6675 0.9996
[2,]  0.1112 0.3963
[3,]  0.9991 0.6527
> cor(X2.als.opa$S, cbind(c(bdata$sp1), c(bdata$sp2)))
        [,1]    [,2]
[1,]  0.6609 0.9998
[2,]  0.9573 0.5535
[3,]  0.9839 0.7053
```

In both cases the third and first MCR component correspond to the true spectra. Clearly, in this case the gain of using two data matrices rather than one is limited: although the spectrum of the second pure compound is estimated better, the first is actually estimated with less accuracy.

This way of combining data matrices also provides an opportunity for quantitation: when some of the components of the mixture are known, one can measure samples in which these components are present in known concentrations. This immediately enables the analyst to convert the area under the curve in the concentration profiles to true concentrations. Additionally, in complex situations such as metabolomics, where many different metabolites are present, some of which having exactly the same UV-Vis spectra because of identical chromophores (parts of the chemical structure leading to absorbance of light at specific wavelengths), it may be extremely difficult to arrive at meaningful solutions (Wehrens et al. 2013). Injecting a couple of standards, or even mixtures of standards, in such cases can really help to disentangle the individual spectra and put the MCR algorithm on the right track. The **alsace** package (Wehrens et al. 2015b) mentioned earlier was designed just for this situation, and provides tools to align and group peaks across samples, split data into several time windows that can be analysed separately, and quantify peak intensities in chromatographic profiles.

References

B.-L. Adam, Y. Qu, J. Davis, M. Ward, M. Clements, L. Cazares, O. Semmes, P. Schellhammer, Y. Yasui, Z. Feng, and G. Wright. Serum protein fingerprinting coupled with a pattern-matching algorithm distinguishes prostate cancer from benign prostate hyperplasia and healthy men. *Cancer Research*, 62:3609–3614, 2002.

H. Akaike. A new look at the statistical model identification. *IEEE Trans. Automatic Control*, 19:716–723, 1974.

V. Badrinarayanan, A. Kendall, and R. Cipolla. Segnet: A deep convolutional encoder-decoder architecture for image segmentation. *IEEE Trans. on Pattern Analysis and Machine Intelligence*, 39:2481–2495, 2017.

J. Banfield and A. Raftery. Model-based Gaussian and non-Gaussian clustering. *Biometrics*, 49:803–821, 1993.

M. Barker and W. Rayens. Partial least squares for discrimination. *J. Chemom.*, 17:166–173, 2003.

I. Barnett, R. Mukherjee, and X. Lin. The generalized higher criticism for testing snp-set effects in genetic association studies. *J. Am. Stat. Assoc.*, 112:64–76, 2017.

A. Barros and D. Rutledge. Genetic algorithms applied to the selection of principal components. *Chemom. Intell. Lab. Syst.*, 40:65–81, 1998.

K. Baumann. Uniform-length molecular descriptors for quantitative structure-property relationships (QSPR) and quantitative structure-activity relationships (QSAR): classification studies and similarity searching. *Trends Anal. Chem.*, 18:36–46, 1999.

K. Baumann, H. Albert, and M. von Korff. A systematic evaluation of the benefits and hazards of variable selection in latent variable regression. Part I. Search algorithm, theory and simulations. *J. Chemom.*, 16:339–350, 2002a.

K. Baumann, H. Albert, and M. von Korff. A systematic evaluation of the benefits and hazards of variable selection in latent variable regression. Part II. Practical applications. *J. Chemom.*, 16:351–360, 2002b.

C. Bergmeir and J. Benítez. Neural networks in R using the Stuttgart Neural Network Simulator: **RSNNS**. *Journal of Statistical Software*, 46:1–26, 2012.

T. Bloemberg, J. Gerretzen, H. Wouters, J. Gloerich, M. van Dael, H. Wessels, L. van den Heuvel, P. Eilers, L. Buydens, and R. Wehrens. Improved parametric time warping for proteomics. *Chemom. Intell. Lab. Systems*, 2010.

I. Borg and P. Groenen. *Modern Multidimensional Scaling*. Springer, 2nd edition, 2005.

A. Boulesteix. PLS dimension reduction for classification with high-dimensional microarray data. *Stat. Appl. Genet. Mol. Biol.*, 3, 2004. Article 33.

© Springer-Verlag GmbH Germany, part of Springer Nature 2020
R. Wehrens, *Chemometrics with R*, Use R!,
https://doi.org/10.1007/978-3-662-62027-4

E. Bouveresse, D. Massart, and P. Dardenne. Modified algorithm for standardization of near-infrared spectrometric instruments. *Anal. Chem.*, 67:1381–1389, 1995.

L. Breiman. Bagging predictors. *Machine Learning*, 24:123–140, 1996.

L. Breiman. Random forests. *Machine Learning*, 45:5–32, 2001.

L. Breiman, J. Friedman, R. Olshen, and C. Stone. *Classification and Regression Trees*. Wadsworth, 1984.

C. Brown and H. Davis. Receiver operating characteristic curves and related decision measures: a tutorial. *Chemom. Intell. Lab. Syst.*, 80:24–38, 2006.

M. Bylesjö, M. Rantalainen, O. Cloarec, J. K. Nicholson, E. Holmes, and J. Trygg. OPLS Discriminant Analysis: combining the strengths of PLS-DA and SIMCA classification. *J. Chemom.*, 20:341–351, 2006.

V. Cerny. A thermodynamical approach to the travelling salesman problem: an efficient simulation algorithm. *Journal of Optimization Theory and Applications*, 45: 41–51, 1985.

S. Challa and R. Potumarthi. Chemometrics-based process analytical technology (pat) tools: Applications and adaptation in pharmaceutical and biopharmaceutical industries. *Applied Biochemistry and Biotechnology*, 169 :66–76, 2013.

A. Chanda, A. M. Daly, D. A. Foley, M. A. LaPack, S. Mukherjee, J. D. Orr, G. L. Reid, D. R. Thompson, and H. W. Ward. Industry perspectives on process analytical technology: Tools and applications in api development. *Organic Process Research & Development*, 19 :63–83, 2015.

W. Chang, J. Cheng, J. Allaire, Y. Xie, and J. McPherson. **shiny***: Web Application Framework for R*, 2018. R package version 1.1.0.

T. Chen and C. Guestrin. Xgboost: A scalable tree boosting system. In *Proceedings of the 22Nd ACM SIGKDD International Conference on Knowledge Discovery and Data Mining*, KDD '16, pages 785–794, New York, NY, USA, 2016. ACM.

H. Chun and S. Keles. Sparse partial least squares for simultaneous dimension reduction and variable selection. *J. Royal Stat. Soc. – Series B*, 72:3–25, 2010.

W. Cleveland. Robust locally weighted regression and smoothing scatterplots. *J. Am. Stat. Assoc.*, 74:829–836, 1979.

D. Clifford and G. Stone. Variable penalty dynamic time warping code for aligning mass spectrometry chromatograms in R. *Journal of Statistical Software*, 47:1–17, 2012.

D. Clifford, G. Stone, I. Montoliu, S. Rezzi, F.-P. Martin, P. Guy, S. Bruce, and S. Kochhar. Alignment using variable penalty dynamic time warping. *Anal. Chem.*, 81:1000–1007, 2009.

T. Cover and J. Thomas. *Elements of Information Theory*. John Wiley & Sons, Ltd., Chichester, 1991.

T. Cox and M. Cox. *Multidimensional Scaling*. Chapman and Hall, 2001.

P. Craven and G. Wahba. Smoothing noisy data with spline functions. *Numer. Math.*, 31:377–403, 1979.

N. Cristianini and J. Shawe-Taylor. *An Introduction to Support Vector Machines and other kernel-based Learning Methods*. Cambridge University Press, 2000.

R. Crosier. Multivariate generalizations of cumulative sum quality-control schemes. *Technometrics*, 30:291–303, 1988.

C. Croux and G. Haesbroeck. Principal components analysis based on robust estimators of the covariance or correlation matrix. *Biometrika*, 87:603–618, 2000.

A. Davison and D. Hinkley. *Bootstrap Methods and their Applications*. Cambridge University Press, Cambridge, 1997.

B. Dayal and J. MacGregor. Improved PLS algorithms. *J. Chemom.*, 11:73–85, 1997.

R. de Gelder, R. Wehrens, and J. Hageman. A generalized expression for the similarity spectra: application to powder diffraction pattern classification. *J. Comput. Chem.*, 22:273–289, 2001.

S. de Jong. SIMPLS: an alternative approach to partial least squares regression. *Chemom. Intell. Lab. Syst.*, 18:251–263, 1993.

S. de Jong and H. Kiers. Principal covariates regression: Part I. Theory. *Chemom. Intell. Lab. Syst.*, 14:155–164, 1992.

A. de Juan, M. Maeder, M. Martinez, and R. Tauler. Combining hard- and soft-modelling techniques to solve kinetic problems. *Chemom. Intell. Lab. Syst.*, 54:49–67, 2000.

O. de Noord. Multivariate calibration standardization. *Chemom. Intell. Lab. Syst.*, 25:85–97, 1994.

A. Dempster, N. Laird, and D. Rubin. Maximum likelihood from incomplete data via the EM algorithm. *J. R. Statist. Soc. B*, 39:1–38, 1977.

L. Deng, J. Li, J. Huang, K. Yao, D. Yu, F. Seide, M. Seltzer, G. Zweig, X. He, J. Williams, Y. Gong, and A. Acero. Recent advances in deep learning for speech research at microsoft. In *2013 IEEE International Conference on Acoustics, Speech and Signal Processing*, pages 8604–8608, 2013.

B. Ding and R. Gentleman. Classification using penalized partial least squares. *J. Comput. Graph. Stat.*, 14:280–298, 2005.

D. Donoho and J. Jin. Higher criticism for detecting sparse heterogeneous mixtures. *The Annals of Statistics*, 32:962–994, 2004.

D. Donoho and J. Jin. Higher criticism thresholding: optimal feature selection when useful features are rare and weak. *Proceedings of the National Academy of Sciences*, 105 :14790–14795, 2008.

S. Dudoit, J. Fridlyand, and T. Speed. Comparison of discrimination methods for the classification of tumors using gene expression data. *J. Am. Stat. Assoc.*, 97:77–87, 2002.

W. Dunn, D. Broadhurst, P. Begley, E. Zelena, S. Francis-McIntyre, N. Anderson, M. Brown, J. Knowles, A. Halsall, J. Haselden, A. Nicholls, I. Wilson, D. Kell, R. Goodacre, and The Human Serum Metabolome (HUSERMET) Consortium. Procedures for large-scale metabolic profiling of serum and plasma using gas chromatography and liquid chromatography coupled to mass spectrometry. *Nat. Protocols*, 6:1060–1083, 2011.

B. Efron. Bootstrap methods: another look at the jackknife. *Ann. Stat.*, 7:1–26, 1979.

B. Efron and T. Hastie. *Computer Age Statistical Inference*. Cambridge University Press, 2016.

B. Efron and R. Tibshirani. *An Introduction to the Bootstrap*. Chapman and Hall, New York, 1993.

B. Efron and R. Tibshirani. Improvements on cross-validation: the .632+ bootstrap method. *J. Am. Stat. Assoc.*, 92:548–560, 1997.

B. Efron, T. Hastie, I. Johnstone, and R. Tibshirani. Least angle regression. *Annals of Statistics*, 32:407–499, 2004.

P. Eilers. Parametric time warping. *Anal. Chem.*, 76:404–411, 2004.

M. Eisen, P. Spellman, P. Brown, and D. Botstein. Cluster analysis and display of genome-wide expression patterns. *Proc. Natl. Acad. Sci. USA*, 95:14863–14868, 1998.

P. Filzmoser, K. Hron, and M. Templ. *Applied Compositional Data Analysis With Worked Examples in R*. Springer Series in Statistics. Springer, 2018.

R. Fisher. The use of multiple measurements in taxonomic problems. *Annals of Eugenics*, 7:179–188, 1936.

M. Forina, C. Armanino, M. Castino, and M. Ubigli. Multivariate data analysis as a discriminating method of the origin of wines. *Vitis*, 25:189–201, 1986.

E. Fowlkes and C. Mallows. A method for comparing two hierarchical clusterings. *J. Am. Stat. Assoc.*, 78:553–584, 1983. Including discussion.

C. Fraley. Algorithms for model-based gaussian hierarchical clustering. *SIAM J. Scient. Comput.*, 20:270–281, 1998.

C. Fraley and A. Raftery. Model-based clustering, discriminant analysis, and density estimation. *J. Am. Stat. Assoc.*, 97:611–631, 2002.

C. Fraley and A. Raftery. Enhanced software for model-based clustering, discriminant analysis, and density estimation: MCLUST. *J. Classif.*, 20:263–286, 2003.

P. Franceschi, D. Masuero, U. Vrhovsek, F. Mattivi, and R. Wehrens. A benchmark spike-in data set for biomarker identification in metabolomics. *J. Chemom.*, 26:16–24, 2012.

I. Frank and J. Friedman. A statistical view of some chemometrics regression tools. *Technometrics*, 35:109–135, 1993.

Y. Freund and R. Schapire. A decision-theoretic generalization of on-line learning and an application to boosting. *J. Comput. Syst. Sci.*, 55:119–139, 1997.

J. Friedman. Exploratory projection pursuit. *Journal of the American Statistical Association*, 82: 249–266, 1987.

J. Friedman. Regularized discriminant analysis. *J. Am. Stat. Assoc.*, 84:165–175, 1989.

J. Friedman. Greedy function approximation: a gradient boosting machine. *Ann. Stat.*, 29:1189–1232, 2001.

J. Friedman and J. Tukey. A projection pursuit algorithm for exploratory data analysis. *IEEE Trans. Comput.*, C23:881–889, 1974.

J. Friedman, T. Hastie, and R. Tibshirani. Additive logistic regression: a statistical view of boosting. *Ann. Stat.*, 28:337–374, 2000.

G. Furnival and G. Wilson. Regression by leaps and bounds. *Technometrics*, 16:499–511, 1974.

K. Gabriel. The biplot graphic display of matrices with application to principal component analysis. *Biometrika*, 58:453–467, 1971.

J. Gagnon-Bartsch and T. Speed. Using control genes to correct for unwanted variation in microarray data. *Biostatistics*, 13:539–552, 2012.

L. A. Gatys, A. S. Ecker, and M. Bethge. Image style transfer using convolutional neural networks. In *2016 IEEE Conference on Computer Vision and Pattern Recognition (CVPR)*, pages 2414–2423, 2016.

P. Geladi, D. MacDougall, and H. Martens. Linearization and scatter-correction for NIR reflectance spectra of meat. *Appl. Spectr.*, 39:491–500, 1985.

T. Giorgino. Computing and visualizing dynamic time warping alignments in R: the dtw package. *J. Stat. Softw.*, 31, 2009.

G. F. Giskeødegård, T. G. Bloemberg, G. Postma, B. Sitter, M.-B. Tessem, I. S. Gribbestad, T. F. Bathen, and L. M. Buydens. Alignment of high resolution magic angle spinning magnetic resonance spectra using warping methods. *Anal. Chim. Acta*, 683:1 – 11, 2010.

D. Goldberg. *Genetic Algorithms in Search, Optimization and Machine Learning*. Kluwer Academic Publishers, Boston, MA., 1989.

I. Goodfellow, Y. Bengio, and A. Courville. *Deep Learning*. MIT Press, 2016. http://www.deeplearningbook.org.

L. Goodman and W. Kruskal. Measures of association for cross classifications. *J. Am. Statist. Assoc.*, 49:732–764, 1954.

J. Gower. Some distance properties of latent root and vector methods used in multivariate analysis. *Biometrika*, 53:325–328, 1966.

J. Gower and D. Hand. *Biplots*. Number 54 in Monographs on Statistics and Applied Probability. Chapman and Hall, London, UK, 1996.

V. Granville, M. Krivanek, and J.-P. Rasson. Simulated annealing: a proof of convergence. *IEEE Trans. Patt. Anal. Machine Intell.*, 16: 652–656, 1994.

F. Günther and S. Fritsch. neuralnet: Training of neural networks. *The R Journal*, 2:30–38, 2010.

Y. Guo, T. Hastie, and R. Tibshirani. Regularized linear discriminant analysis and its application in microarrays. *Biostatistics*, 8:86–100, 2007.

D. Hand and K. Yu. Idiot's Bayes – not so stupid after all? *Int. Statist. Rev.*, 69:385–398, 2001.

W. Härdle and L. Simar. *Applied Multivariate Statistical Analysis*. Springer, Berlin, 2nd edition, 2007.

K. Hasegawa, T. Kimura, Y. Miyashita, and K. Funatsu. Nonlinear partial least squares modeling of phenyl alkylamines with the monoamine oxidase inhibitory activities. *J. Chem. Inf. Comput. Sci-*, 36:1025–1029, 1996.

T. Hastie, R. Tibshirani, and J. Friedman. *The Elements of Statistical Learning*. Springer Series in Statistics. Springer, New York, 2001.

M. Hendriks, F. van Eeuwijk, R. Jellema, J. Westerhuis, T. Reijmers, H. Hoefsloot, and A. Smilde. Data-processing strategies for metabolomics studies. *Tr. Anal. Chem.*, 30:1685–1698, 2011.

G. Hinton, L. Deng, D. Yu, G. E. Dahl, A. Mohamed, N. Jaitly, A. Senior, V. Vanhoucke, P. Nguyen, T. N. Sainath, and B. Kingsbury. Deep neural networks for acoustic modeling in speech recognition: The shared views of four research groups. *IEEE Signal Processing Magazine*, 29: 82–97, 2012.

A. Hoerl. Application of ridge analysis to regression problems. *Chemical Engineering Progress*, 58:54–59, 1962.

A. Hoerl and R. Kennard. Ridge regression: biased estimation for non-orthogonal problems. *Technometrics*, 8:27–51, 1970.

A. Hoerl, R. Kennard, and K. Baldwin. Ridge regression: some simulations. *Commun. Stat. – Simul. Comput.*, 4:105–123, 1975.

J. Honaker, G. King, and M. Blackwell. Amelia II: A program for missing data. *Journal of Statistical Software*, 45:1–47, 2011.

P. Huber. Projection pursuit. *The Annals of Statistics*, 13:435–475, 1985.

L. Hubert. Comparing partitions. *J. Classif.*, 2:193–218, 1985.

M. Hubert. Robust calibration. In S. Brown, R. Tauler, and B. Walczak, editors, *Comprehensive Chemometrics – Chemical and Biochemical Data Analysis*, chapter 3.07, pages 315–343. Elsevier, 2009.

M. Hubert and K. V. Branden. Robust methods for partial least squares regression. *J. Chemom.*, 17:537–549, 2003.

M. Hubert, P. Rousseeuw, and K. V. Branden. ROBPCA: a new approach to robust principal component analysis. *Technometrics*, 47:64–79, 2005.

R. Huo, R. Wehrens, J. van Duynhoven, and L. Buydens. Assessment of techniques for DOSY NMR data processing. *Anal. Chim. Acta*, 490:231–251, 2003.

R. Huo, R. Wehrens, and L. Buydens. Improved DOSY NMR data processing by data enhancement and combination of multivariate curve resolution with non-linear least squares fitting. *J. Magn. Reson.*, 169:257–269, 2004.

A. Hyvärinen and E. Oja. Independent component analysis: algorithms and applications. *Neural Networks*, 13:411–430, 2000.

A. Hyvärinen, J. Karhunen, and E. Oja. *Independent Component Analysis*. John Wiley & Sons, Ltd., Chichester, 2001.

J. Jackson. *A User's Guide to Principal Components*. John Wiley & Sons, Ltd., Chichester, 1991.

I. Jolliffe. *Principal Component Analysis*. Springer, New York, 1986.

K. Jorgensen, V. H. Segtnan, K. Thyholt, and T. Næs. A comparison of methods for analysing regression models with both spectral and designed variables. *J. Chemom.*, 18:451–464, 2004.

J. Josse and F. Husson. missMDA: A package for handling missing values in multivariate data analysis. *Journal of Statistical Software*, 70:1–31, 2016.

J. Josse, J. Pagès, and F. Husson. Multiple imputation in PCA. *Adv. Data Anal. Classif.*, 5:231–246, 2011.

J. Kalivas. Two data sets of near infrared spectra. *Chemom. Intell. Lab. Syst*, 37:255–259, 1997.

L. Kaufman and P. Rousseeuw. *Finding Groups in Data – An Introduction to Cluster Analysis*. John Wiley & Sons, New York, 1990.

H. Keller and D. Massart. Peak purity control in liquid chromatography with photodiode-array detection by a fixed size moving window evolving factor analysis. *Anal. Chim. Acta*, 246:379–390, 1991.

R. Kennard and L. Stone. Computer aided design of experiments. *Technometrics*, 11:137–148, 1969.

S. Kirkpatrick, C. Gelatt, and M. P. Vecchi. Optimization by simulated annealing. *Science*, 220:671–680, 1983.

K. Kjeldahl and R. Bro. Some common misunderstandings in chemometrics. *J. Chemom.*, 24:558–564, 2010.

T. Kohonen. *Self-Organizing Maps*. Number 30 in Springer Series in Information Sciences. Springer, Berlin, 3 edition, 2001.

T. Kourti. Application of latent variable methods to process control and multivariate statistical process control in industry. *International Journal of Adaptive Control and Signal Processing*, 19:213–246, 2005.

T. Kourti and J. F. MacGregor. Process analysis, monitoring and diagnosis, using multivariate projection methods. *Chemometrics and Intelligent Laboratory Systems*, 28 :3 – 21, 1995.

N. Krämer, A.-L. Boulesteix, and G. Tutz. Penalized partial least squares with applications to B-spline transformations and functional data. *Chemom. Intell. Lab. Syst.*, 94:60–69, 2008.

M. Kuhn. Building predictive models in r using the caret package. *Journal of Statistical Software, Articles*, 28 :1–26, 2008.

M. Kuhn and K. Johnson. *Applied Predictive Modeling*. Springer, 2013.

A. Land and A. Doig. An automatic method for solving discrete programming problems. *Econometrica*, 28:497–520, 1960.

J. Lawless and P. Wang. A simulation study of ridge and other regression estimators. *Commun. Stat. – Theory and Methods*, 5:303–323, 1976.

R. D. Lawrence, G. S. Almasi, and H. E. Rushmeier. A scalable parallel algorithm for self-organizing maps with applications to sparse data mining problems. *Data Mining and Knowledge Discovery*, 3: 171–195, 1999.

W. Lawton and E. Sylvestre. Self-modeling curve resolution. *Technometrics*, 13:617–633, 1971.

R. Leardi. Genetic algorithms in chemometrics and chemistry: a review. *J. Chemom.*, 15:559–569, 2001.

B. Liebmann, P. Filzmoser, and K. Varmuza. Robust and classical PLS regression compared. *J. Chemom.*, 24:111–120, 2009.

R. Little and D. Rubin. *Statistical Analysis with Missing Data*. Wiley Series in Probability and Statistics. John Wiley & Sons, Inc., 3rd edition, 2019.

A. D. Livera, M. Sysi-Aho, L. Jacob, J. Gagnon-Bartsch, S. Castillo, J. Simpson, and T. P. Speed. Statistical methods for handling unwanted variation in metabolomics data. *Anal. Chem.*, 87:3606–3615, 2015.

M. Maeder. Evolving factor analysis for the resolution of overlapping chromatographic peaks. *Anal. Chem.*, 59:527–530, 1987.

C. Mallows. Some comments on Cp. *Technometrics*, 15:661–675, 1973.

K. Mardia, J. Kent, and J. Bibby. *Multivariate Analysis*. Academic Press, 1979.

R. Maronna, R. Martin, and V. Yohai. *Robust Statistics: Theory and Methods*. Wiley Series in Probability and Statistics. John Wiley & Sons, Ltd., 2005.

H. Martens and T. Næs. *Multivariate Calibration*. John Wiley & Sons, Ltd., Chichester, 1989.

B. Marx. Iteratively reweighted partial least squares estimation for generalized linear regression. *Technometrics*, 38:374–381, 1996.

G. McLachlan. *Discriminant Analysis and Statistical Pattern Recognition*. Wiley-Interscience, 2004.

G. McLachlan and T. Krishnan. *The EM Algorithm and Extensions*. John Wiley & Sons, 1997.

G. McLachlan and D. Peel. *Finite Mixture Models*. John Wiley & Sons, New York, 2000.

M. Meila. Comparing clusterings – an information-based distance. *J. Multivar. Anal.*, 98:873–895, 2007.

N. Meinshausen and P. Bühlmann. Stability selection. *Journal of the Royal Statistical Society B*, 72: 417–473, 2010. With discussion.

N. Metropolis, A. Rosenbluth, M. Rosenbluth, A. Teller, and E. Teller. Equations of state calculations by fast computing machines. *J. Chem. Phys.*, 21:1087–1092, 1953.

B.-H. Mevik and H. Cederkvist. Mean squared error of prediction (MSEP) estimates for principal component regression (PCR) and partial least squares regression (PLSR). *J. Chemom.*, 18:422–429, 2004.

B.-H. Mevik and R. Wehrens. The pls package: principal component and partial least squares regression in R. *J. Stat. Soft.*, 18, 2007.

G. Michailides, K. Johnson, and M. Culp. ada: an R package for stochastic boosting. *J. Stat. Softw.*, 17, 2006.

D. Montgomery. *Introduction To Statistical Quality Control*. John Wiley and Sons, 4th ed. edition, 2001.

K. Mullen, I. van Stokkum, and V. Mihaleva. Global analysis of multiple gas chromatography-mass spectrometry (GC/MS) data sets: a method for resolution of co-eluting components with comparison to MCR-ALS. *Chemom. Intell. Lab. Syst.*, 95:150–163, 2009.

G. Munoz and A. de Juan. pH- and time-dependent hemoglobin transitions: a case study for process modelling. *Anal. Chim. Acta*, 595:198–208, 2007.

T. Næs, T. Isaksson, and B. Kowalski. Locally weighted regression and scatter correction for near-infrared reflectance data. *Anal. Chem.*, 62:664–673, 1990.

G. Nason. *Wavelet methods in statistics with R.* Springer, New York, 2008.

A. B. Nassif, I. Shahin, I. Attili, M. Azzeh, and K. Shaalan. Speech recognition using deep neural networks: A systematic review. *IEEE Access*, 7:19143–19165, 2019.

D. Nguyen and D. Rocke. Tumor classification by partial least squares using microarray gene expression data. *Bioinformatics*, 18:39–50, 2002.

A. Niazi and R. Leardi. Genetic algorithms in chemometrics. *J. Chemom.*, 26:345–351, 2012.

N. Nielsen, J. Carstensen, and J. Smedsgaard. Aligning of single and multiple wavelength chromatographic profiles for chemometric data analysis using correlation optimized warping. *J. Chrom. A*, 805:17–35, 1998.

M. Olteanu and N. Villa-Vialaneix. On-line relational and multiple relational som. *Neurocomputing*, 147:15–30, 2015.

E. Page. Cumulative sum charts. *Technometrics*, 3:1–9, 1961.

J. Pignatiello and G. Runger. Comparisons of multivariate CUSUM charts. *J. Qual. Tech.*, 22:173–186, 1990.

Y. Qu, B.-L. Adam, Y. Yasui, M. Ward, L. Cazares, P. Schellhammer, Z. Feng, O. Semmes, and G. Wright. Boosted decision tree analysis of surface-enhanced laser desorption/ionization mass spectral serum profiles discriminates prostate cancer from noncancer patients. *Clin Chem*, 48:1835–43, 2002.

J. Quinlan. The induction of decision trees. *Mach. Learn.*, 1:81–106, 1986.

J. Quinlan. *C4.5: Programs for Machine Learning.* Morgan Kaufmann, 1993.

R Development Core Team. *R: A Language and Environment for Statistical Computing.* R Foundation for Statistical Computing, Vienna, Austria, 2010. ISBN 3-900051-07-0.

L. Rabiner, A. Rosenberg, and S. Levinson. Considerations in dynamic time warping algorithms for discrete word recognition. *IEEE Trans. Acoust., Speech, Signal Process.*, 26: 575–582, 1978.

W. Rand. Objective criteria for the evaluation of clustering methods. *J. Am. Stat. Assoc.*, 66:846–850, 1971.

S. Rännar, F. Lindgren, P. Geladi, and S. Wold. The PLS Kernel algorithm for data sets with many variables and fewer objects. Part 1: Theory and algorithm. *J. Chemom.*, 8:111, 1994.

B. Ripley. *Pattern recognition and neural networks.* Cambridge University Press, 1996.

D. Risso, J. Ngai, T. Speed, and S. Dudoit. Normalization of RNA-seq data using factor analysis of control genes or samples. *Nat. Biotechnology*, 32:896, 2014.

S. Roberts. Control chart tests based on geometric moving averages. *Technometrics*, 42:97–102, 1959.

F. Rosenblatt. *Principles of neurodynamics.* Spartan Books, Washington DC, 1962.

P. Rousseeuw. Least median of squares regression. *J. Am. Stat. Assoc.*, 79:871–880, 1984.

P. Rousseeuw and W. V. den Bossche. Detecting deviating data cells. *Technometrics*, 60:135–145, 2018.

P. Rousseeuw and K. van Driessen. A fast algorithm for the minimum covariance determinant estimator. *Technometrics*, 41:212–223, 1999.

D. Rumelhard and J. McClelland, editors. *Parallel distributed processing: explorations in the microstructure of cognition. Volume 1: Foundations.* MIT Press, Cambridge MA, 1986.

O. Russakovsky, J. Deng, H. Su, J. Krause, S. Satheesh, S. Ma, Z. Huang, A. Karpathy, A. Khosla, M. Bernstein, A. C. Berg, and L. Fei-Fei. ImageNet Large Scale Visual Recognition Challenge. *International Journal of Computer Vision (IJCV)*, 115 :211–252, 2015.

H. Sakoe and S. Chiba. Dynamic programming algorithm optimization for spoken word recognition. *IEEE Trans. Acoust., Speech, Signal Process.*, 26: 43–49, 1978.

F. C. Sanchez, B. van de Bogaert, S. Rutan, and D. Massart. Multivariate peak purity approaches. *Chemom. Intell. Lab. Syst.*, 36:153–164, 1996.

F. Q. Sanchez, M. Khots, D. Massart, and J. de Beer. Algorithm for the assessment of peak purity in liquid chromatography with photodiode-array detection. *Anal. Chim. Acta*, 285:181–192, 1994.

E. Santos-Fernández. *Multivariate Statistical Quality Control Using R*, volume 14. Springer, 2013.

A. Savitsky and M. Golay. Smoothing and differentiation of data by simplified least squares procedures. *Anal. Chem.*, 36:1627–1639, 1964.

R. Schapire, Y. Freund, P. Bartlett, and W. Lee. Boosting the margin: a new explanation for the effectiveness of voting methods. *Ann. Stat.*, 26:1651–1686, 1998.

B. Schölkopf and A. Smola. *Learning with kernels*. MIT Press, Cambridge, MA, 2002.

G. Schwarz. Estimating the dimension of a model. *Ann. Statist.*, 6:461–464, 1978.

L. Scrucca. GA: A package for genetic algorithms in R. *Journal of Statistical Software*, 53:1–37, 2013.

L. Scrucca. On some extensions to GA package: hybrid optimisation, parallelisation and islands evolution. *The R Journal*, 9:187–206, 2017.

J. Shao. Linear model selection by cross-validation. *J. Am. Statist. Assoc.*, 88:486–494, 2003.

A. D. Silva. Efficient variable screening for multivariate analysis. *J. Mult. Anal.*, 76:35–62, 2001.

S. Smit, M. van Breemen, H. Hoefsloot, A. Smilde, J. Aerts, and C. de Koster. Assessing the statistical validity of proteomics based biomarkers. *Anal. Chim. Acta*, 592:210–217, 2007.

C. Spearman. "General intelligence", objectively determined and measured. *Am. J. Psychol.*, 15:201–293, 1904.

P. Stilbs. Molecular self-diffusion coefficients in Fourier transform nuclear magnetic resonance spectrometric analysis of complex mixtures. *Anal. Chem.*, 53:2135–2137, 1981.

P. Stilbs. Fourier-transform pulsed-gradient spin-echo studies of molecular diffusion. *Progr. NMR Spectrosc.*, 19:1–45, 1987.

M. Stone. Cross-validatory choice and assessment of statistical predictions. *J. R. Statist. Soc. B*, 36:111–147, 1974. Including discussion.

M. Stone and R. Brooks. Continuum regression: cross-validated sequentially constructed prediction embracing ordinary least squares, partial least squares and principal components regression (with discussion). *J. R. Statist. Soc.*, 52:237–269, 1990.

O. Svensson, T. Kourti, and J. MacGregor. A comparison of orthogonal signal correction algorithms and characteristics. *J. Chemom.*, 16:176–188, 2002.

V. Svetnik, A. Liaw, C. Tong, J. Culberson, R. Sheridan, and B. Feuston. Random forest: a classification and regression tool for compound classification and qsar modeling. *J. Chem. Inf. Comput. Sci.*, 43:1947–58, 2003.

H. Tapp and E. Kemsley. Notes on the practical utility of OPLS. *Tr. Anal. Chem.*, 28:1322–1327, 2009.

R. Tauler. Multivariate curve resolution applied to second-order data. *Chemom. Intell. Lab. Syst.*, 30:133–146, 1995.

R. Tauler, S. Lacorte, and D. Barceló. Application of multivariate self-modeling curve resolution to the quantitation of trace levels of organophosphorus pesticides in natural waters from interlaboratory studies. *J. Chromatogr. A*, 730:177–183, 1996.

M. Tenenhaus, V. E. Vinzi, Y. Chatelin, and C. Lauro. PLS path modelling. *Comput. Stat. Data Anal.*, 48:159–2005, 2005.

C. ter Braak and S. Juggins. Weighted averaging partial least squares regression WAPLS: an improved method for reconstructing environmental variables from species assemblages. *Hydrobiologia*, 269:485–502, 1993.

T. Therneau and E. Atkinson. An introduction to recursive partitioning using the RPART routines. Technical Report 61, Mayo Foundation, 1997.

E. A. Thévenot, A. Roux, Y. Xu, E. Ezan, and C. Junot. Analysis of the human adult urinary metabolome variations with age, body mass index, and gender by implementing a comprehensive workflow for univariate and opls statistical analyses. *Journal of Proteome Research*, 14: 3322–3335, 2015.

U. Thissen, H. Swierenga, A. de Weijer, R. Wehrens, W. Melssen, and L. Buydens. Multivariate statistical process control using mixture modelling. *J. Chemom.*, 19:23–31, 2005.

R. Tibshirani. Regression shrinkage and selection via the lasso. *J. Royal. Statist. Soc B*, 58:267–288, 1996.

R. Tibshirani, T. Hastie, B. Narashimhan, and G. Chu. Class prediction by nearest shrunken centroids with applications to dna microarrays. *Statistical Science*, 18:104–117, 2003.

V. Todorov and P. Filzmoser. An object oriented framework for robust multivariate analysis. *J. Stat. Softw.*, 32:1–47, 2009.

J. Trygg and S. Wold. Orthogonal projections to latent structures (O-PLS). *J. Chemom.*, 16:119–128, 2002.

H. Uh, F. Hartgers, M. Yazdankakhsh, and J. Houwing-Duistermaat. Evaluation of regression methods when immunological measurements are constrained by detection limits. *BMC Immunology*, 9, 2008.

J. R. R. Uijlings, K. E. A. van de Sande, T. Gevers, and A. W. M. Smeulders. Selective search for object recognition. *International Journal of Computer Vision*, 104 :154–171, 2013.

A. Ultsch. Self-organizing neural networks for visualization and classification. In O. Opitz, B. Lausen, and R. Klar, editors, *Information and Classification – Concepts, Methods and Applications*, pages 307–313. Springer Verlag, 1993.

S. van Buuren and K. Groothuis-Oudshoorn. mice: Multivariate imputation by chained equations in r. *Journal of Statistical Software*, 45:1–67, 2011.

H. Van der Voet. Comparing the predictive accuracy of models using a simple randomization test. *Chemom. Intell. Lab. Syst.*, 25:313–323, 1994.

V. Vapnik. *The Nature of Statistical Learning Theory*. Springer-Verlag, 1995.

K. Varmuza and P. Filzmoser. *Introduction to Multivariate Statistical Analysis in Chemometrics*. Taylor & Francis - CRC Press, Boca Raton, FL, USA, 2009.

W. Venables, D. Smith, and the R development Core Team. An introduction to R, 2009. Version 2.10.1.

W. N. Venables and B. D. Ripley. *Modern Applied Statistics with S*. Springer, New York, fourth edition, 2002. ISBN 0-387-95457-0.

J. Vesanto and E. Alhoniemi. Clustering of the self-organising map. *IEEE Trans. Neural Netw.*, 11:586–600, 2000.

C. Wang and T. Isenhour. Time-warping algorithm applied to chromatographic peak matching gas chromatography / Fourier Transform infrared / Mass Spectrometry. *Anal. Chem.*, 59:649–654, 1987.

Y. Wang, M. Lysaght, and B. Kowalski. Improvement of multivariate calibration through instrument standardization. *Anal. Chem.*, 64:764–771, 1992.

Y. Wang, D. Veltkamp, and B. Kowalski. Multivariate instrument standardization. *Anal. Chem.*, 63:2750–2756, 1994.

Z. Wang, T. Dean, and B. Kowalski. Additive background correction in multivariate instrument standardization. *Anal. Chem.*, 67:2379–2385, 1995.

R. Wehrens and L. Buydens. Self- and super-organising maps in R: the kohonen package. *Journal of Statistical Software*, 21, 2007.

R. Wehrens and P. Franceschi. Meta-statistics for variable selection: The R package BioMark. *Journal of Statistical Software*, 51:1–18, 2012a.

R. Wehrens and P. Franceschi. Thresholding for biomarker selection using Higher Criticism. *Mol. Biosystems*, 9:2339–2346, 2012b.

R. Wehrens and J. Kruisselbrink. Flexible self-organising maps in kohonen v3.0. *Journal of Statistical Software*, 87, 2018.

R. Wehrens and W. van der Linden. Bootstrapping principal-component regression models. *J. Chemom.*, 11:157–171, 1997.

R. Wehrens and E. Willighagen. Mapping databases of x-ray powder patterns. *R News*, 6:24–28, 2006.

R. Wehrens, P. Franceschi, U. Vrhovsek, and F. Mattivi. Stability-based biomarker selection. *Anal. Chim. Acta*, 705:15–23, 2011.

R. Wehrens, E. Carvalho, D. Masuero, A. de Juan, and S. Martens. High-throughput carotenoid profiling using multivariate curve resolution. *Anal. Bioanal. Chem.*, 15:5075–5086, 2013.

R. Wehrens, T. Bloemberg, and P. Eilers. Fast parametric time warping of peak lists. *Bioinformatics*, 31:3063–3065, 2015a.

R. Wehrens, E. Carvalho, and P. Fraser. Metabolite profiling in LC-DAD using multivariate curve resolution: the alsace package for R. *Metabolomics*, 11:143–154, 2015b.

R. Wehrens, J. A. Hageman, F. van Eeuwijk, R. Kooke, P. J. Flood, E. Wijnker, J. J. Keurentjes, A. Lommen, H. D. van Eekelen, R. D. Hall, R. Mumm, and R. C. de Vos. Improved batch correction in untargeted MS-based metabolomics. *Metabolomics*, 12:1–12, 2016.

J. A. Westerhuis, S. P. Gurden, and A. K. Smilde. Generalized contribution plots in multivariate statistical process monitoring. *Chemometrics and Intelligent Laboratory Systems*, 51 :95 – 114, 2000.

Western Electric Co. *Statistical Quality Control Handbook*. Western Electric Co., Indianapolis, Indiana, 1st edition, 1956.

W. Windig and J. Guilment. Interactive self-modeling mixture analysis. *Anal. Chem.*, 63:1425–1432, 1991.

W. Windig, J. Phalp, and A. Payna. A noise and background reduction method for component detection in liquid chromatography/mass spectrometry. *Anal. Chem.*, 68:3602–3606, 1996.

P. Wittek, S. Gao, I. Lim, and L. Zhao. **somoclu**: An efficient parallel library for self-organizing maps. *Journal of Statistical Software*, 78:1–21, 2017.

S. Wold, N. Kettaneh-Wold, and B. Skagerberg. Nonlinear PLS modeling. *Chemom. Intell. Lab. Syst.*, 7:53–65, 1989.

S. Wold, H. Antti, F. Lindgren, and J. Ohman. Orthogonal signal correction of near-infrared spectra. *Chemom. Intell. Lab. Syst.*, 44:175–185, 1998.

B. Worley and R. Powers. Multivariate analysis in metabolomics. *Curr. Metabolomics*, 1:92–107, 2013.

Y. Xie. knitr: A comprehensive tool for reproducible research in R. In V. Stodden, F. Leisch, and R. D. Peng, editors, *Implementing Reproducible Computational Research*. Chapman and Hall/CRC, 2014. ISBN 978-1466561595.

Y. Xie. *Dynamic Documents with R and knitr*. Chapman and Hall/CRC, Boca Raton, Florida, 2nd edition, 2015. ISBN 978-1498716963.

Y. Xie. *knitr: A General-Purpose Package for Dynamic Report Generation in R*, 2018. R package version 1.20.

J. Yan. **som**: *Self-Organizing Map*, 2016. R package version 0.3-5.1.

H. Zou and T. Hastie. Regularization and variable selection via the elastic net. *J. Royal. Stat. Soc. B*, 67:301–320, 2005.

Index

© Springer-Verlag GmbH Germany, part of Springer Nature 2020
R. Wehrens, *Chemometrics with R*, Use R!,
https://doi.org/10.1007/978-3-662-62027-4

Printed in the United States
By Bookmasters